FOREWORD

The Bridge Pressure Flow Scour for Clear Water Conditions Study described in this report was conducted at the Federal Highway Administration's (FHWA) Turner-Fairbank Highway Research Center (TFHRC) J. Sterling Jones Hydraulics Laboratory. The study was in response to a request of several State transportation departments asking for a new design guidance to predict bridge pressure flow scour for clear water conditions. The new pressure flow scour procedure will replace the existing pressure flow scour prediction method in the FHWA Hydraulic Engineering Circular No. 18 (4th edition) *Evaluating Scour at Bridges*. The study includes experiments (physical modeling) at the Hydraulics Laboratory. This report will be of interest to hydraulic and bridge engineers who are involved in estimating pressure flow scour for inundated bridge decks. This report is being distributed as an electronic document through the TFHRC Web site (www.tfhrc.gov).

Cheryl Allen Richter
Acting Director, Office of Infrastructure
Research and Development

This document is disseminated under the sponsorship of the U.S. Department of Transportation in the interest of information exchange. The U.S. Government assumes no liability for its contents or use thereof. This report does not constitute a standard, specification, policy, or regulation.

The U.S. Government does not endorse products or manufacturers. Trade and manufacturers' names appear in this report only because they are considered essential to the object of the document.

Quality Assurance Statement

The Federal Highway Administration (FHWA) provides high-quality information to serve Government, industry, and the public in a manner that promotes public understanding. Standards and policies are used to ensure and maximize the quality, objectivity, utility, and integrity of its information. FHWA periodically reviews quality issues and adjusts its programs and processes to ensure continuous quality improvement.

TECHNICAL REPORT DOCUMENTATION PAGE

1. Report No. FHWA-HRT-09-041	2. Government Accession No.	3. Recipient's Catalog No. N/A
4. Title and Subtitle Bridge Pressure Flow Scour for Clear Water Conditions		5. Report Date October 2009
		6. Performing Organization Code N/A
7. Author(s) Junke Guo, Kornel Kerenyi, and Jorge E. Pagan-Ortiz		8. Performing Organization Report No.
9. Performing Organizations Names and Addresses GKY and Associates, Inc. University of Nebraska 4229 Lafayette Center Dr. 312 N. 14th Street Suite 1850 Alexander Building West Chantilly, VA 20151 Lincoln, NE 68588-0430		10. Work Unit No. (TRAIS) N/A
		11. Contract or Grant No.
12. Sponsoring Agency Name and Address Office of Infrastructure Research and Development Federal Highway Administration 6300 Georgetown Pike McLean, VA 22101-2296		13. Type of Report and Period Covered Laboratory Report
		14. Sponsoring Agency Code

15. Supplementary Notes
The Contracting Officer's Technical Representative (COTR) was Kornel Kerenyi, HRDI-07. Oscar Berrios assisted with experimentation and produced some of the figures. Kevin Flora, Denis Lyn, and Bart Bergendahl provided constructive suggestions.

16. Abstract
The equilibrium scour at a bridge caused by pressure flow with critical approach velocity in clear-water simulation conditions was studied both analytically and experimentally. The flume experiments revealed that (1) the measured equilibrium scour profiles under a bridge are more or less consistent across the channel width; (2) all the measured scour profiles can be described by two similarity equations where the horizontal distance is scaled by the deck width and the local scour is scaled by the maximum scour depth; (3) the maximum scour position is located under the bridge and at a location approximately 15.4 percent of the deck width from the downstream edge of the deck; (4) scour begins at approximately one deck width upstream of the bridge, and deposition begins at approximately 2.5 deck widths downstream of the bridge; and (5) the maximum scour depth decreases with increasing median sediment size but increases with higher levels of deck inundation. The analytical analysis shows that (1) bridge scour can be divided into three cases: downstream unsubmerged, partially submerged, and totally submerged; (2) for downstream unsubmerged flows, the maximum scour depth is an open channel problem where the conventional methods in terms of critical velocity or bed shear stress can be applied; and (3) for partially and totally submerged flows, the maximum scour depth can be described by scour and inundation similarity numbers, which has been confirmed by experiments with two sediment sizes (0.039 and 0.078 inches (1 and 2 mm)) and two types of decks with three and six girders, respectively. For application, a design and field evaluation procedure with examples is presented, including the maximum scour depth and scour profile.

17. Key Words Bridge decks, Bridge design, Bridge foundations, Bridge hydraulics, Bridge inundation, Bridge scour, Pressure flows, Pressure scour, Submerged flows	18. Distribution Statement No restrictions. This document is available to the public through the National Technical Information Service (NTIS), Springfield, VA 22161.		
19. Security Classif. (of this report) Unclassified	20. Security Classif. (of this page) Unclassified	21. No. of Pages 57	22. Price

Form DOT F 1700.7 (8-72)   Reproduction of completed page authorized.

SI* (MODERN METRIC) CONVERSION FACTORS

APPROXIMATE CONVERSIONS TO SI UNITS

Symbol	When You Know	Multiply By	To Find	Symbol
LENGTH				
in	inches	25.4	millimeters	mm
ft	feet	0.305	meters	m
yd	yards	0.914	meters	m
mi	miles	1.61	kilometers	km
AREA				
in^2	square inches	645.2	square millimeters	mm^2
ft^2	square feet	0.093	square meters	m^2
yd^2	square yard	0.836	square meters	m^2
ac	acres	0.405	hectares	ha
mi^2	square miles	2.59	square kilometers	km^2
VOLUME				
fl oz	fluid ounces	29.57	milliliters	mL
gal	gallons	3.785	liters	L
ft^3	cubic feet	0.028	cubic meters	m^3
yd^3	cubic yards	0.765	cubic meters	m^3
NOTE: volumes greater than 1000 L shall be shown in m^3				
MASS				
oz	ounces	28.35	grams	g
lb	pounds	0.454	kilograms	kg
T	short tons (2000 lb)	0.907	megagrams (or "metric ton")	Mg (or "t")
TEMPERATURE (exact degrees)				
°F	Fahrenheit	5 (F-32)/9 or (F-32)/1.8	Celsius	°C
ILLUMINATION				
fc	foot-candles	10.76	lux	lx
fl	foot-Lamberts	3.426	candela/m^2	cd/m^2
FORCE and PRESSURE or STRESS				
lbf	poundforce	4.45	newtons	N
lbf/in^2	poundforce per square inch	6.89	kilopascals	kPa

APPROXIMATE CONVERSIONS FROM SI UNITS

Symbol	When You Know	Multiply By	To Find	Symbol
LENGTH				
mm	millimeters	0.039	inches	in
m	meters	3.28	feet	ft
m	meters	1.09	yards	yd
km	kilometers	0.621	miles	mi
AREA				
mm^2	square millimeters	0.0016	square inches	in^2
m^2	square meters	10.764	square feet	ft^2
m^2	square meters	1.195	square yards	yd^2
ha	hectares	2.47	acres	ac
km^2	square kilometers	0.386	square miles	mi^2
VOLUME				
mL	milliliters	0.034	fluid ounces	fl oz
L	liters	0.264	gallons	gal
m^3	cubic meters	35.314	cubic feet	ft^3
m^3	cubic meters	1.307	cubic yards	yd^3
MASS				
g	grams	0.035	ounces	oz
kg	kilograms	2.202	pounds	lb
Mg (or "t")	megagrams (or "metric ton")	1.103	short tons (2000 lb)	T
TEMPERATURE (exact degrees)				
°C	Celsius	1.8C+32	Fahrenheit	°F
ILLUMINATION				
lx	lux	0.0929	foot-candles	fc
cd/m^2	candela/m^2	0.2919	foot-Lamberts	fl
FORCE and PRESSURE or STRESS				
N	newtons	0.225	poundforce	lbf
kPa	kilopascals	0.145	poundforce per square inch	lbf/in^2

*SI is the symbol for the International System of Units. Appropriate rounding should be made to comply with Section 4 of ASTM E380.
(Revised March 2003)

TABLE OF CONTENTS

LIST OF FIGURES

LIST OF TABLES

LIST OF ABBREVIATIONS AND SYMBOLS

Abbreviations

2D	Two-dimensional
3D	Three-dimensional
FHWA	Federal Highway Administration
TFHRC	Turner-Fairbank Highway Research Center

Symbols

a	Deck block depth
b	Thickness of bridge deck including girders
B	Width of a river
d_*	Dimensionless sediment diameter
d_{50}	Median diameter of sediment
F	Inundation Froude number
Fr	Froude number
g	Gravitational acceleration
h	Downstream flow depth in case 1
h_b	Bridge opening
h_d	Bridge downstream flow depth
h_u	Depth of headwater
K_b	Bridge energy loss coefficient
K_p	Curvature pressure coefficient
K_s	Critical Shields number
m	Fitting parameter in the bridge energy loss coefficient
n	Manning coefficient, or normal direction of a streamline
p_1	Pressure at point 1
p_2	Pressure at point 2
Q	Operating discharge in the flume
q	Unit discharge of a river
q_1	Unit discharge through the bridge
$\mathbf{R}$	Local radius of curvature of a streamline
R_0	Radius of curvature at the maximum scour point
R^2	Correlation coefficient
Re	Reynolds number
s	Specific gravity of sediment
v	Kinematic viscosity of water
V_a	Velocity through the bridge before scour

V_b	Velocity through the bridge at the maximum scour depth
V_c	Critical velocity
V_u	Velocity of the headwater
V_{uc}	Upstream critical velocity
V_{ue}	Upstream effective velocity
W	Width of bridge
x	Coordinate along a river
x_1	Coordinate of upstream face of deck
x_2	Coordinate of downstream face of deck
x_d	Coordinate of initiation of deposition
x_s	Coordinate of initiation of scour
y_s	Maximum scour depth
z	Vertical direction
α_1, α_2	Energy correction coefficients
β	Correction factor for hydrostatic pressure under bridge
λ	An empirical fitting factor
γ	Specific weight of water
τ_c	Critical shear stress

CHAPTER 1. INTRODUCTION

Bridges are a vital component of the transportation network. Evaluating their stability and structural response after a flood event is critical to highway safety. Bridge studies are usually designed with an assumption of an open channel flow condition, but the flow regime can switch to pressure flow when the downstream edge of a bridge deck is partially or totally submerged during a large flood. Figure 1 shows a bridge undergoing partially submerged flow in Salt Creek, NE, in June 2008. Figure 2 shows a totally submerged flow in Cedar River, IA, in June 2008, which interrupted traffic on I-80. Unlike open channel flows, these pressure flows create a severe scourability potential because scouring the channel bed is one of the only ways to dissipate the energy when passing a given discharge in pressurized flow.

Although most bridge scour events are due to live bed scour, a maximum scour depth often results from clear water flows with a critical approach velocity for bedload motion. For bridge safety, this report emphasizes the equilibrium maximum scour of pressure flows in extreme clear water conditions.

The objectives of the study were to collect a detailed high-quality dataset of pressure flow scour at a model bridge and to develop an analytical solution for pressure flow scour based on mass and energy conservation laws. To these ends, existing results in the literature were reviewed, and knowledge gaps were identified. Next, a series of flume experiments were conducted to examine the existing methods and test new hypotheses on bridge pressure flow scour. After, bridge flows were divided into three cases, and the mass and energy conservation laws were applied to each case, leading to hypotheses for pressure flow scour predictions. The hypotheses were tested with the flume data. In this report, an example procedure for calculating the maximum scour depth and scour profile is presented along with recommended research needs.

Figure 1. Photo. Partially inundated bridge deck at Salt Creek, NE.

Source: Iowa Department of Transportation
(Photo provided by Keven Arrowsmith)

Figure 2. Photo. Completely inundated bridges at Cedar River, IA.

CHAPTER 2. LITERATURE REVIEW

To better understand pressure flow scour, three systematic studies were completed by Arneson and Abt, Lyn, and Umbrell et al.[1–3] Arneson and Abt conducted a series of flume tests and proposed the following regression equation in figure 3:

$$\frac{y_s}{h_u} = -0.93 + 0.23\left(\frac{h_u}{h_b}\right) + 0.82\left(\frac{y_s + h_b}{h_u}\right) + 0.03\left(\frac{V_a}{V_{uc}}\right)$$

Figure 3. Equation. Arneson and Abt's scour depth equation.[1]

Where:

y_s = The maximum scour depth.
h_u = The depth of the headwater.
h_b = The vertical bridge opening at the main channel before scouring.
V_a = The velocity through the bridge before scour.
V_{uc} = The upstream critical velocity, as defined in the equation in figure 4.

$$V_{uc} = 1.52\sqrt{g(s-1)d_{50}}\left(\frac{h_u}{d_{50}}\right)^{1/6}$$

Figure 4. Equation. Upstream critical velocity.

Where:

g = The gravitational acceleration.
s = The specific gravity of sediment.
d_{50} = The median diameter of the bed materials.

Although the equation in figure 3 has been adopted in the FHWA manual, it presents a serious problem.[4] As Lyn states, the equation in figure 3 suffers from a spurious correlation since both sides of the equation include the term y_s/h_u.[2] As an alternative, Lyn proposes the following power law equation for scour depth in figure 5:

$$\frac{y_s}{h_u} = \min\left[0.105\left(\frac{V_a}{V_{uc}}\right)^{2.95}, 0.5\right]$$

Figure 5. Equation. Lyn's scour depth equation.[2]

The third study was conducted by Umbrell et al., who performed a series of flume experiments at the TFHRC Hydraulics Laboratory in McLean, VA.[3] Using the law of conservation of mass and assuming that the velocity under the bridge at equilibrium scour is approximately equal to the critical velocity of the upstream flow, they developed the equation as shown in figure 6:

$$\frac{y_s + h_b}{h_u} = \frac{V_u}{V_{uc}}\left(1 - \frac{b}{h_u}\right)$$

Figure 6. Equation. Umbrell et al.'s scour depth equation.[3]

Where:

V_u = The velocity of the headwater.
b = The thickness of the bridge deck including girders.

By comparing figure 6 with their experimental data, Umbrell et al. modified it to generate the equation in figure 7 as follows:

$$\frac{y_s + h_b}{h_u} = 1.102\left[\frac{V_u}{V_{uc}}\left(1 - \frac{b}{h_u}\right)\right]^{0.603}$$

Figure 7. Equation. Modified Umbrell et al. scour depth equation.[3]

V_{uc} is calculated as in figure 4, except the coefficient 1.52 is replaced by 1.58. The equation in figure 7 raises three concerns. First, the under-bridge V_{uc} was not necessarily the same as that upstream. Furthermore, the inclusion of b invalidated the equation for partially submerged flows because the flow depth would only rise to a position on the bridge deck with a height less than b. Last, the tests were run for only 3.5 hours, which was not enough time for equilibrium scour to develop. The tests performed for this report, which were conducted in the same flume that Umbrell et al. used, showed that equilibrium scour was attained after 32–48 hours. For a detailed review of the Arneson and Abt and Umbrell et al. datasets, refer to the recent paper by Lyn, which questions the quality of the two datasets.[2]

In summary, the study of bridge pressure flow scour was not developed sufficiently to be useful in bridge design. The two primary datasets were not of high quality and did not have information on the characteristics of scour profiles. The three analyses were empirical and lacked a theoretical explanation for the mechanism of pressure flow scour. To advance the study of pressure flow scour, it is important to acquire new firsthand data. Thus, a series of experiments are introduced in the following chapter.

CHAPTER 3. EXPERIMENTAL STUDY

The objective of the experiments was to collect scour data at a bridge under controlled pressure flow conditions in a flume. The collected data was then used to formulate a general understanding of bridge pressure flow scour and to test both the existing prediction equations and a new hypothesis proposed in this report. To this end, a series of flume tests were conducted in the Hydraulics Laboratory. The experimental setup, results, data analysis, and interpretation are described in the following subsections.

EXPERIMENTAL SETUP

Flume System

Figure 8 shows an overview of the experimental flume, and 1 m = 2.38 ft

figure 9 details the flume system. The flume is a rectangle and 70.03 ft (21.35 m) long by 6 ft (1.83 m) wide. It has glass sides and a stainless steel bottom. In the middle of the flume, a test section 2.07 ft (0.63 m) wide and 9.18 ft (2.8 m) long was installed, and a model bridge deck was mounted within as seen in 1 m = 2.38 ft

figure 9. A honeycomb flow straightener and a trumpet-shaped inlet were designed to smoothly guide the flow into the test channel. Referring to the side view in 1 m = 2.38 ft

figure 9, a 15.60-inch (40-cm) sediment recess was installed along the flume bottom and under the bridge to record local scour information. The flume was set horizontally, and an adjustable tailgate located at the downstream end of the flume controlled the depth of flow.

A circulation system with a sump and a pump supplied water to the flume. The capacity of the sump was 7,415.94 ft^3 (210 m^3), and the pump output rate varied between 0 and 10.59 ft^3/s (0 and 0.3 m^3/s). An electromagnetic flowmeter was used to measure the discharge. More information about the flume can be found at http://www.fhwa.dot.gov/engineering/hydraulics/research/lab.cfm.

Figure 8. Photo. Approach section of the test flume.

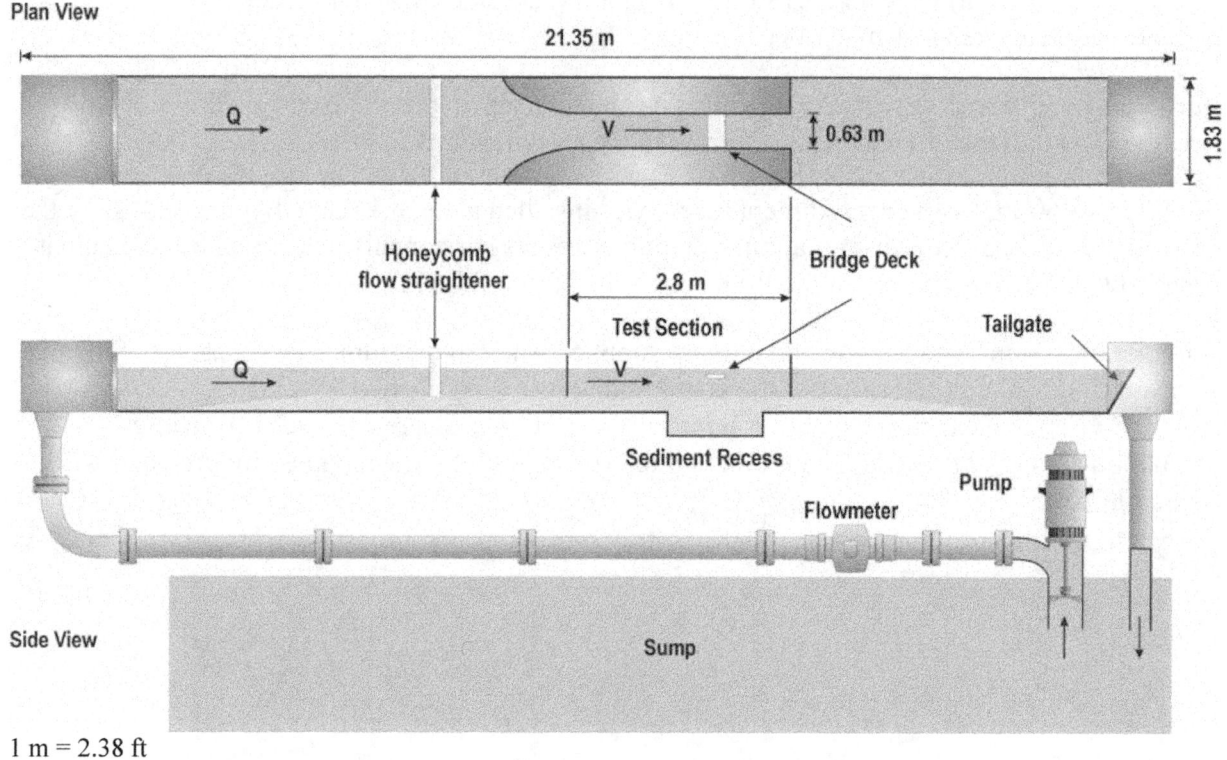

Figure 9. Illustration. Plan and side schematic of the test flume.

Sand Bed Preparation

Figure 10 shows the sand bed preparation in the test channel. The sand diameter was roughly uniform, and $d_{50} = 0.039$ inches (1 mm). A 7.80-inch (20-cm)-thick layer of sand was distributed evenly on the flume bottom. The sediment recess under the bridge was deep enough to model a local scour to a depth of 23.40 inches (60 cm). To test the effect of sediment size, sand with

$d_{50} = 0.078$ inches (2 mm) was also used.

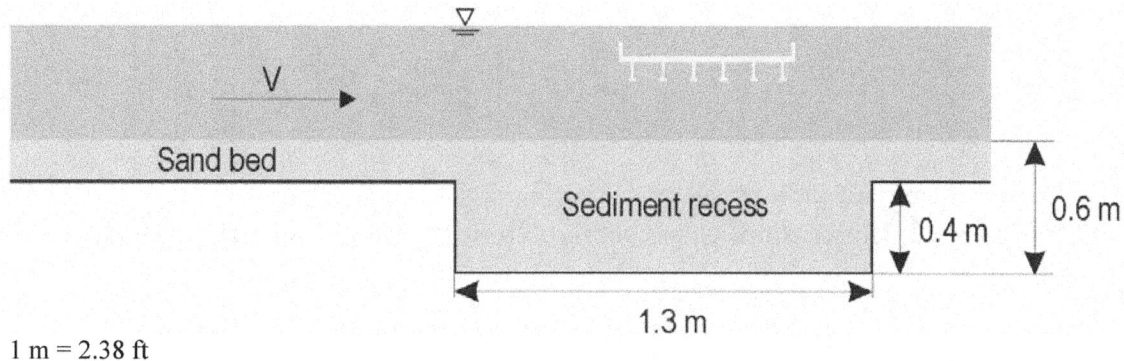

1 m = 2.38 ft

Figure 10. Illustration. Detail of sand bed and sediment recess in test flume.

Model Decks

Two Plexiglas® model decks with three girders and six girders were used in the tests, and they are shown in figure 11 through figure 13. The six-girder deck was chosen since most four-lane U.S. highway bridges have six girders, while the three-girder deck is more common for two-lane bridges. As shown in figure 12 and figure 13, both decks had rails at the edges and had the same width of 0.85 ft (0.26 m), though they did not have the same height. Figure 11 more clearly shows the spaces in the railing that allowed flow to pass in a three-dimensional (3D) view. The deck elevation was adjustable, permitting the deck to have eight bridge opening heights.

Figure 11. Illustration. 3D view of a six-girder bridge deck.

7

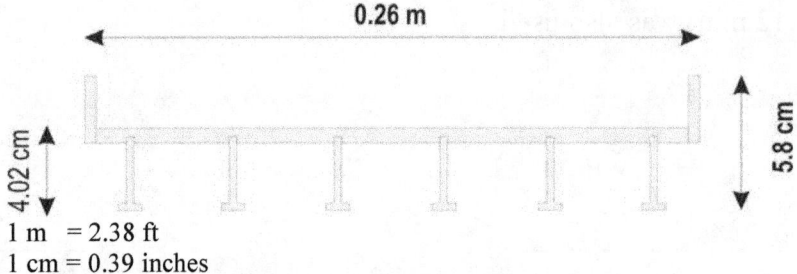

Figure 12. Illustration. Cross section view of a six-girder bridge deck.

Figure 13. Illustration. Cross section view of a three-girder bridge deck.

Operating Discharge

To ensure a maximum clear water scour under the bridge, the approach velocity in the test section should have been at critical velocity for bedload initiation. Since the flow depth was always kept at 9.75 inches (25 cm) during the experiments, the critical velocity for $d_{50} = 0.039$ inches (1 mm) was $V_{uc} = 1.41$ ft/s (0.43 m/s) according to the method proposed by Neill.[5] The upstream approach velocity was then chosen as $V_u = 0.95 \times V_{uc} = 1.34$ ft/s (0.41 m/s), which resulted in an operating discharge, Q, estimated in the following equation in figure 14:

$$Q = V_u \cdot B \cdot h_u = (0.41 \, \text{m/s})(0.63 \, \text{m})(0.25 \, \text{m}) = 0.0646 \, \text{m}^3 / \text{s}$$

Figure 14. Equation. Operating discharge Q.

Where:

B = The width of the test section, 2.07 ft (0.63 m).
h_u = The flow depth, 0.82 ft (0.25 m).

The Reynolds number is $Re = V_u h_u / v = 1.025 \times 10^5$, where $v =$ kinematic viscosity of water, and the Froude number $Fr = V_u / (g h_u)^{1/2} = 0.26$.

Similarly, for $d_{50} = 0.078$ inches (2 mm), V_{uc} was estimated to be 1.84 ft/s (0.56 m/s). V_u was then chosen as $0.95 \times V_{uc} = 1.74$ ft/s (0.53 m/s), which corresponded to a discharge of 2.9487 ft^3/s (0.0835 m^3/s) with $Re = 1.325 \times 10^5$ and $Fr = 0.34$.

Data Collection

An automated flume carriage fitted to the main flume, seen in figure 15, was used to collect scour data that were measured using a laser distance sensor. A LabVIEW™ virtual instrument was programmed for data acquisition, instrument control, data analysis, and report generation.

**Figure 15. Photo. Automated flume carriage with laser distance sensor
perched over the test flume.**

Experimental Procedure

The following steps outline the experimental procedure:

1. The sand bed was installed, as shown in figure 10.

2. A bridge deck was installed and positioned perpendicular to the direction of flow.

3. The elevation of the deck was adjusted to a designated bridge opening.

4. Water was gradually pumped from the sump to the flume until the operating discharge was met, and it was verified with the electromagnetic flowmeter.

5. Each experiment was run for 32–48 hours until the equilibrium scour state was attained.

6. The 3D scour hole was mapped using the laser distance sensor.

EXPERIMENTAL RESULTS

The major experimental results were based on the 3D scour mapping recordings from the sand recess. They are presented in 3D visualizations, longitudinal profiles, and maximum scour depth.

Figure 16 represents a 3D scour hole, showing a more-or-less uniform scour perpendicular to the direction of flow. This means that the scour holes could be approximated across the entire width of the test section by a two-dimensional (2D) scour profile in the longitudinal dimension. Figure 17 through figure 19 present plots of all of the width-averaged scour profiles, where x = zero is at the maximum scour point that is 1.56 inches (4 cm) from the downstream deck edge, and y = zero is the elevation at the top of the sand bed before scour. The figures show that the scour profiles were roughly bell-shaped curves, but they were not symmetrical because the eroded materials deposited about two to three times the deck width downstream the bridge, where y > zero. In addition, the scour began at about 1 deck width upstream of the bridge, and the scour decreased with increasing sediment size, though it is noted that the approach velocity in figure 19 was larger than that in figure 18. The maximum scour depths are tabulated in column 2 of table 1 through table 3, showing that y_s increased as h_b in column 1 decreased or as the bridge inundation increased.

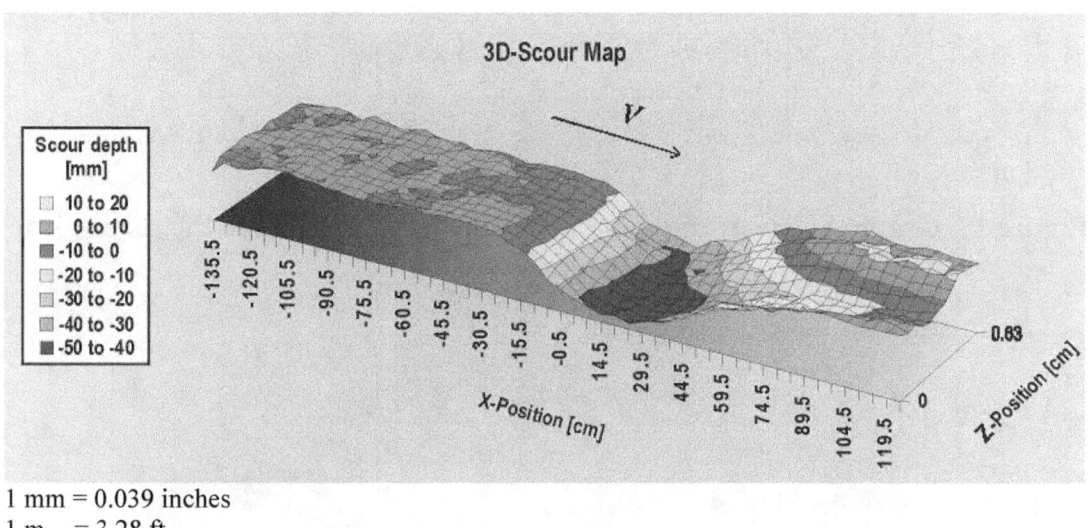

1 mm = 0.039 inches
1 m   = 3.28 ft

Figure 16. Graph. 3D scour map at equilibrium scour.

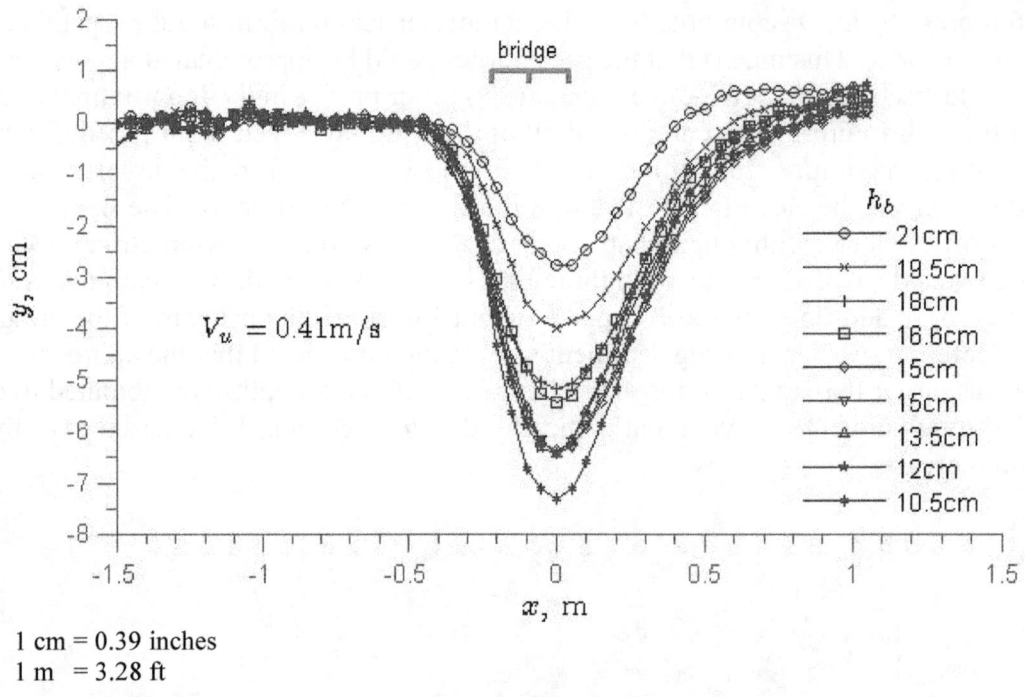

1 cm = 0.39 inches
1 m = 3.28 ft

Figure 17. Graph. Scour profiles at various bridge openings for the three-girder bridge deck.

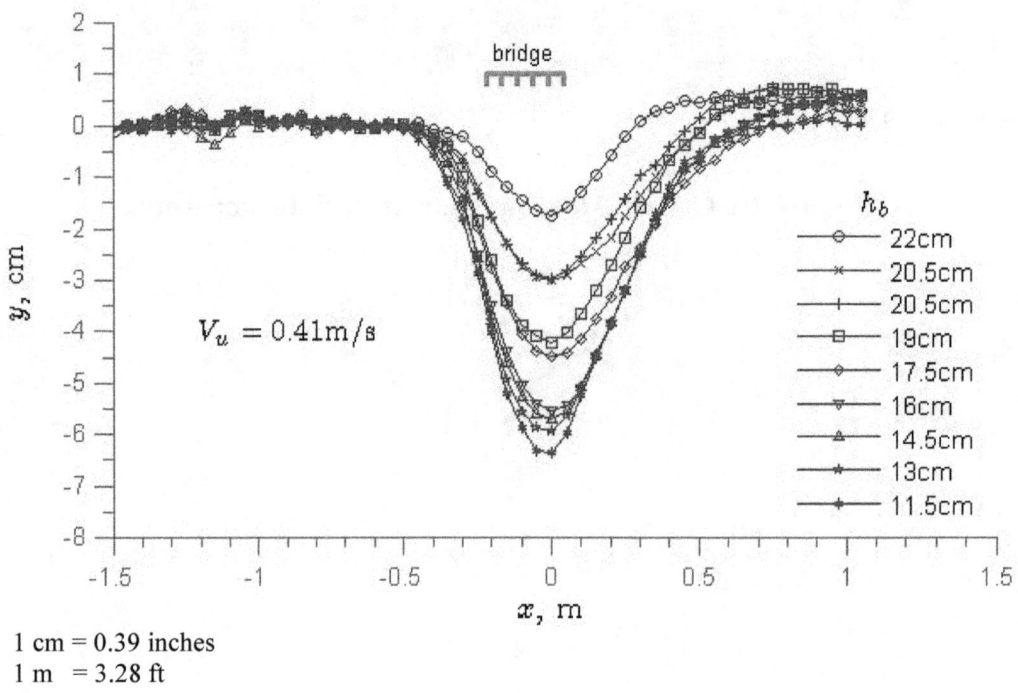

1 cm = 0.39 inches
1 m = 3.28 ft

Figure 18. Graph. Scour profiles at various bridge openings for the six-girder bridge deck.

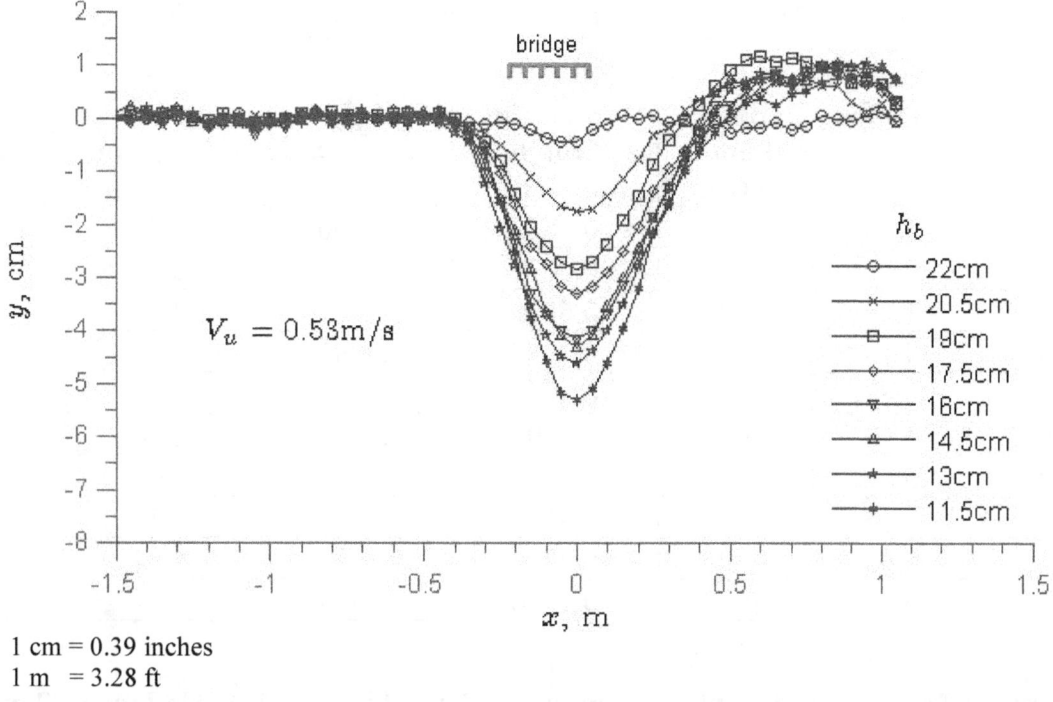

1 cm = 0.39 inches
1 m = 3.28 ft

Figure 19. Graph. Scour profiles at various bridge openings for the six-girder bridge deck (d_{50} = 0.078 inches (2 mm)).

Table 1. Experimental results for the three-girder bridge (d_{50} = 0.039 inches (1 mm)).

Bridge Opening, h_b (cm)	Measured Scour Depth, y_s (cm)	Block Depth, a (cm)	Inundation Froude number, F	Scour Number, $\dfrac{h_b + y_s}{h_b + a}$	Calculated Scour Depth, y_s (cm)	Relative Error (Percent)
21.0	2.77	4.00	0.6545	0.9508	2.67	-3.7
19.5	3.98	5.06	0.5499	0.9559	4.38	10.2
18.0	5.18	5.06	0.4620	1.0051	5.12	-1.1
16.5	5.45	5.06	0.3959	1.0180	5.76	5.7
15.0	6.35	5.06	0.3433	1.0642	6.30	-0.7
15.0	6.42	5.06	0.3433	1.0677	6.30	-1.8
13.5	6.41	5.06	0.2997	1.0726	6.76	5.5
12.0	6.43	5.06	0.2624	1.0802	7.14	11.0
10.5	7.31	5.06	0.2298	1.1444	7.44	1.8

1 cm = 0.39 inches

13

Table 2. Experimental results for the six-girder bridge ($d_{50} = 0.039$ inches (1 mm)).

Bridge Opening, h_b (cm)	Measured Scour Depth, y_s (cm)	Block Depth, a (cm)	Inundation Froude number, F	Scour Number, $\dfrac{h_b + y_s}{h_b + a}$	Calculated Scour Depth, y_s (cm)	Relative Error (Percent)
22.0	1.75	3.00	0.7558	0.9500	1.24	-29.2
20.5	2.99	4.02	0.6070	0.9580	2.97	-0.6
20.5	2.98	4.02	0.6070	0.9575	2.97	-0.3
19.0	4.23	4.02	0.4982	1.0091	3.77	-10.9
19.0	4.52	4.02	0.4982	1.0217	3.77	-16.6
17.5	4.47	4.02	0.4208	1.0209	4.46	-0.2
16.0	5.55	4.02	0.3613	1.0764	5.05	-9.1
14.5	5.71	4.02	0.3131	1.0912	5.54	-3.0
13.0	5.93	4.02	0.2726	1.1122	5.94	0.2
11.5	6.34	4.02	0.2376	1.1494	6.27	-1.1

1 cm = 0.39 inches

Table 3. Experimental results for the six-girder bridge, ($d_{50} = 0.078$ inches (2 mm)).

Bridge Opening, h_b (cm)	Measured Scour Depth, y_s (cm)	Block Depth, a (cm)	Inundation Froude number, F	Scour Number, $\dfrac{h_b + y_s}{h_b + a}$	Calculated Scour Depth, y_s (cm)	Relative Error (Percent)
22.0	0.43	3.00	0.9770	0.8973	0.75	74.0
20.5	1.75	4.02	0.7847	0.9074	2.20	25.8
19.0	2.83	4.02	0.6441	0.9484	2.84	0.3
17.5	3.29	4.02	0.5440	0.9661	3.46	5.2
16.0	4.14	4.02	0.4670	1.0060	4.03	-2.6
14.5	4.30	4.02	0.4047	1.0151	4.54	5.6
13.0	4.62	4.02	0.3523	1.0350	4.98	7.9
11.5	5.31	4.02	0.3071	1.0829	5.36	0.9
7.0	6.50	4.02	0.1988	1.2249	6.09	-6.3
2.5	11.64	4.02	0.1138	2.1684	6.20	-46.8

1 cm = 0.39 inches

DATA ANALYSIS AND INTERPRETATION

Similarity of Scour Profiles

By looking at all the profiles in figure 17 through figure 19, it is hypothesized that a similarity profile may exist by normalizing x to the deck width, W, and y to y_s. This hypothesis was tested in figure 20 and figure 21, and the figures demonstrate a surprising similarity for $x \leq$ zero, corresponding to the pressure flow scour. The scatter for $x >$ zero was due to the influence of the downstream free surface.

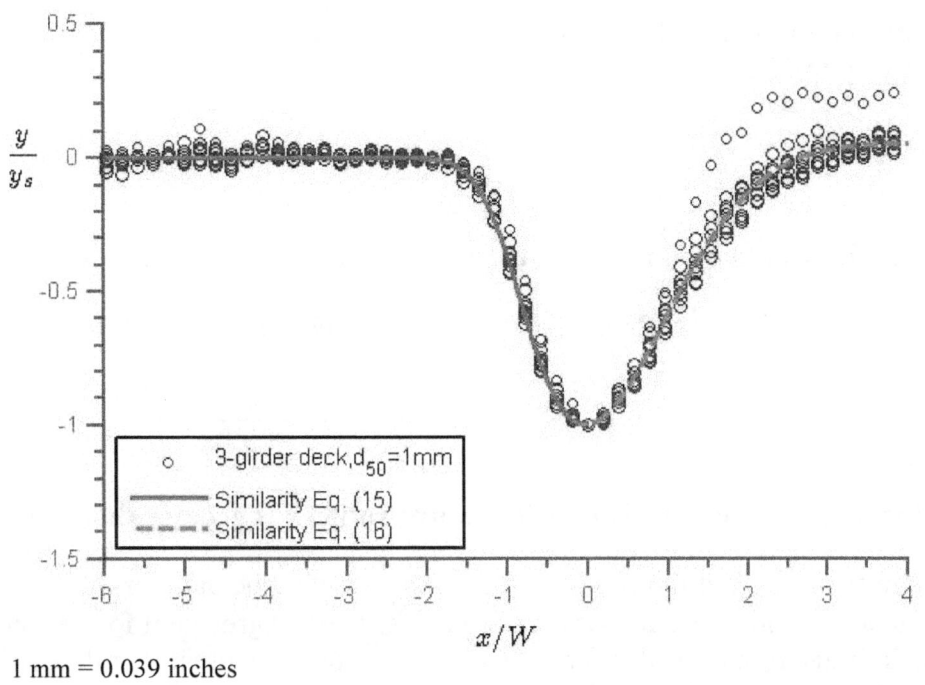

1 mm = 0.039 inches

Figure 20. Graph. Similarity profile for equilibrium scour for the three-girder bridge deck.

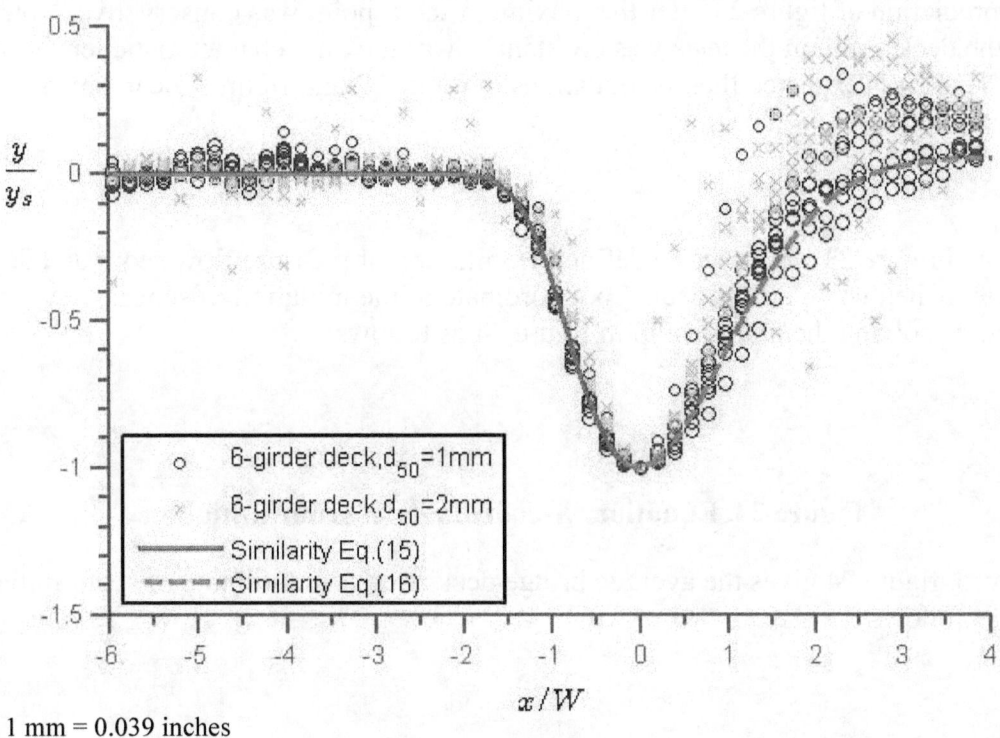

1 mm = 0.039 inches

Figure 21. Graph. Similarity profile for equilibrium scour for the six-girder bridge deck.

For the scour profiles in figure 20 under the three-girder deck, the similarity profile for $x \leq$ zero is arranged into figure 22 as follows:

$$\frac{y}{y_s} = -\exp\left(-\left|\frac{x}{W}\right|^{2.5}\right)$$

Figure 22. Equation. Similarity scour profile, x is less than or equal to zero.

For $x >$ zero, the profile in figure 20 can be approximated by figure 23 as follows:

$$\frac{y}{y_s} = -1.055\exp\left[-\frac{1}{2}\left(\frac{x}{W}\right)^{1.8}\right] + 0.055$$

Figure 23. Equation. Similarity scour profile, x is greater than zero.

The equations in figure 22 and figure 23 were also plotted with the data for the six-girder deck with two sediment sizes in figure 21, which showed very good agreement for $x \leq$ zero but an overestimation of most scour profiles for $x >$ zero (i.e., scour was greater or deposition was less than the experimental results). Figure 22 describes a pressure flow scour profile before the maximum scour point well, which is independent of the number of girders and sediment size. Also, the prediction of figure 23 after the maximum scour point was conservative. Note that although the deck width in the tests was constant, it was the only characteristic length in the flow direction. Thus, it is expected that the equations in figure 22 and figure 23 can be applied to other deck widths.

INTERPRETATION

Figure 22 and figure 23 were used to define the initiation of pressure flow scour and deposition. Scour started when $y/y_s = -0.01$, and the x-coordinate of the initiation of scour, x_s, was then determined by solving the relationship in figure 24 as follows:

$$-0.01 = -\exp\left(-\left|\frac{x_s}{W}\right|^{2.5}\right)$$

Figure 24. Equation. X-coordinate of scour initiation.

When solved, figure 24 gives the average bridge deck width x-coordinate of scour initiation in figure 25 as follows:

$$\frac{x_s}{W} = -1.842$$

Figure 25. Equation. X-coordinate of scour initiation normalized to bridge width.

The dimensional abscissa, x_1, of the upstream face of the bridge is found according to figure 26 as follows:

$$x_1 = -(26\,cm - 4\,cm) = -22\,cm$$

Figure 26. Equation. Upstream dimensional abscissa, x_1.

The width, W, of the model deck was 10.14 inches (26 cm), the distance between the maximum scour depth and the downstream face of the bridge was 1.56 inches (4 cm), and the minus sign meant the scour began before the maximum scour depth. The dimensionless abscissa, x_1/W, of the upstream face of the bridge was then expressed as shown in figure 27:

$$\frac{x_1}{W} = -\frac{22\,cm}{26\,cm} = -0.84615$$

Figure 27. Equation. Dimensionless abscissa upstream.

The relative distance between the scour initiation and the upstream deck face is solved in figure 28, which means the scour begins at about 1 deck width upstream the bridge.

$$\frac{x_1}{W} - \frac{x_s}{W} = -0.84615 - (-1.842) = 0.99585 \approx 1$$

Figure 28. Equation. Distance from scour initiation position to bridge deck face.

The deposition position x_d can be defined by y/y_s = zero in figure 23, which gives the following equation in figure 29:

$$\frac{x_d}{W} = 2.6827$$

Figure 29. Equation. Initiation of sediment deposition position.

Considering the dimensionless abscissa, x_2/W, of the downstream deck face was 0.154 (see figure 30), the relative distance between the downstream deck face and the deposition point was 2.53 (see figure 31). This means the deposition began at a distance of about 2.5 times the deck width downstream of the bridge, as shown in figure 31.

$$\frac{x_2}{W} = \frac{4\,cm}{26\,cm} = 0.154$$

Figure 30. Equation. Dimensionless abscissa downstream.

$$\frac{x_d}{W} - \frac{x_2}{W} = 2.6827 - 0.15385 = 2.53$$

Figure 31. Equation. Distance from bridge deck to deposition position.

Similarly, figure 22 and figure 23 gave the relative scour depths at the two deck edges, which were useful for field scour evaluation and will be detailed with an example later in the report. The equation in figure 32 gave the normalized depth at the deck edges. For applications, figure 33 gave a normalized scour profile with highlighted positions of interest. For reference, the bridge deck is shown in the figure.

$$\left.\frac{y}{y_s}\right|_{\text{upstream deck edge}} = -0.518, \text{ and } \left.\frac{y}{y_s}\right|_{\text{downstream deck edge}} = -0.985$$

Figure 32. Equation. Scour depth at deck edges.

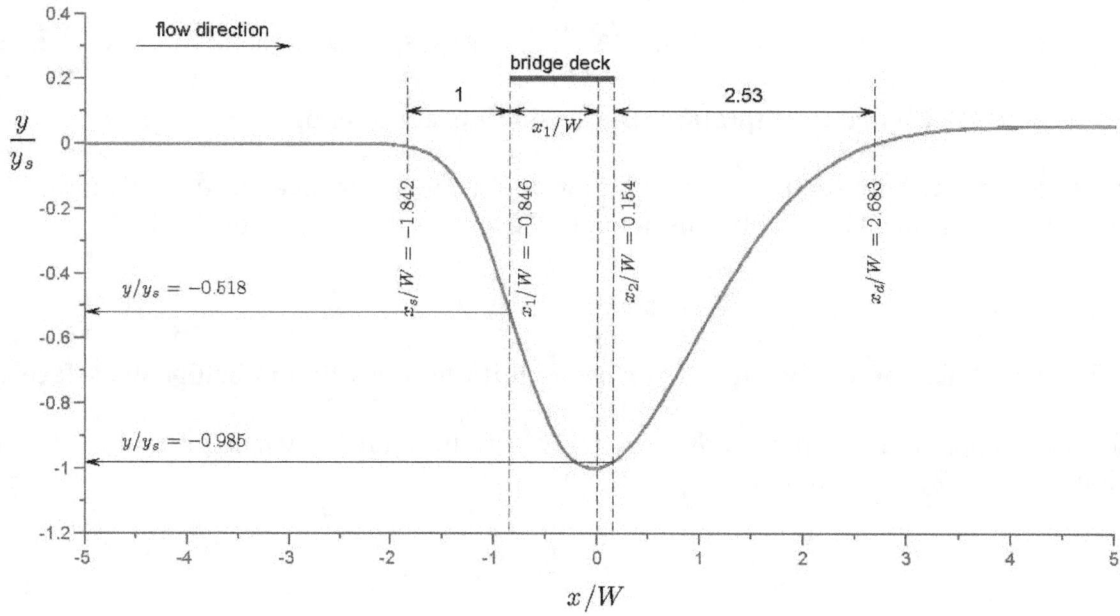

Figure 33. Graph. Normalized scour profile.

In short, the horizontal scour domain of bridge under a pressure flow condition depended on the width of the bridge deck, but the design of a scour profile by figure 22 and figure 23 needed the maximum y_s.

DETERMINATION OF THE MAXIMUM SCOUR DEPTH USING THE EXISTING METHODS

The three methods mentioned in the literature review were tested with the current data in figure 34 through 1 mm = 0.039 inches
1 m   = 3.28 ft

figure 36 in which the overflow had been subtracted according to Umbrell et al.[3] The Arneson and Abt method had an inverse correlation with the test data, which means the functional structure of figure 3 was not correct. Lyn's method underestimated most of the present data (see figure 5). Although the Umbrell et al. method is the best of the existing methods in terms of application, none of them provide reliable predictions (see figure 6). In next chapter, an analytical method is provided for estimating the maximum y_s by applying the mass and energy conservation laws.

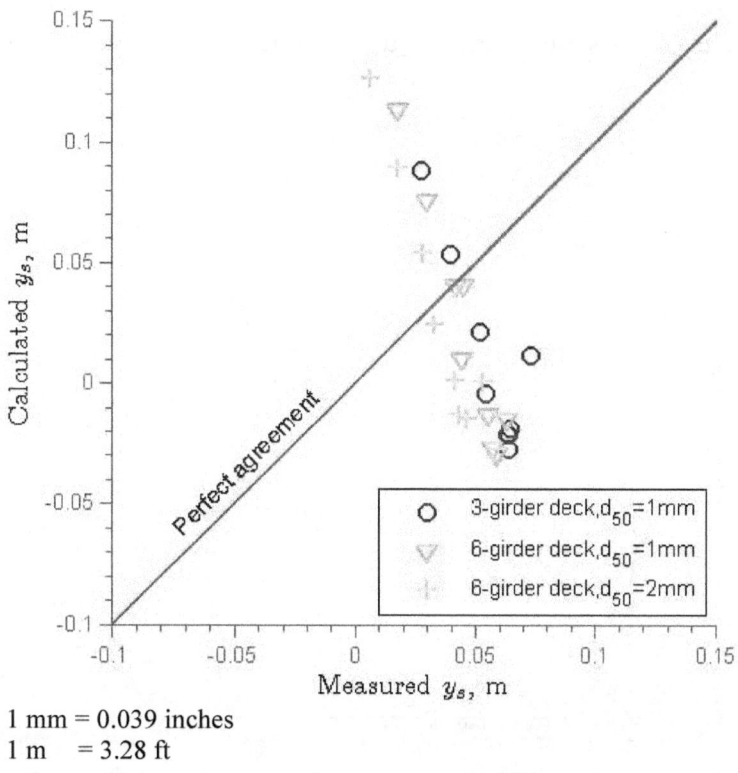

1 mm = 0.039 inches
1 m   = 3.28 ft

Figure 34. Graph. Arneson and Abt's scour depth equation agreement with experimental data.[1]

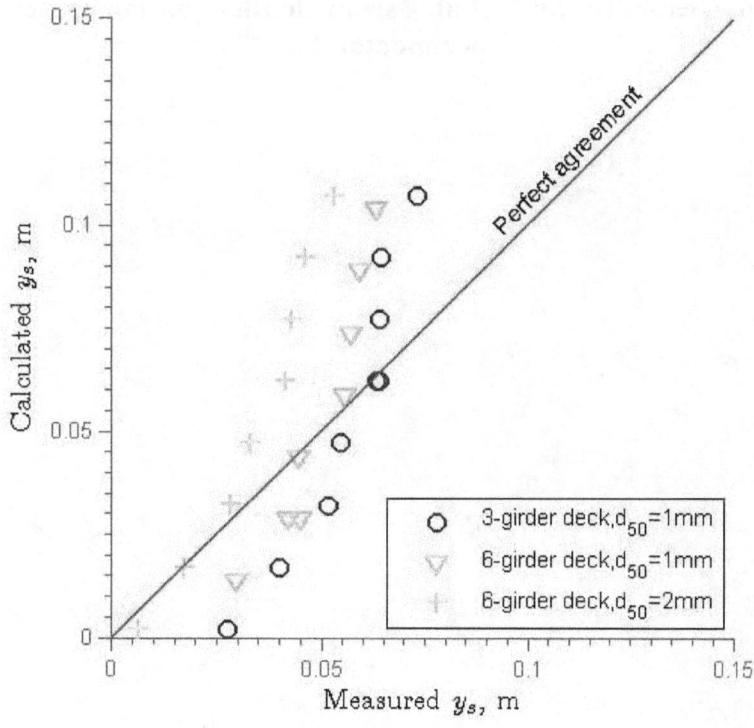

Figure 35. Graph. Lyn's scour depth equation agreement with experimental data.[2]

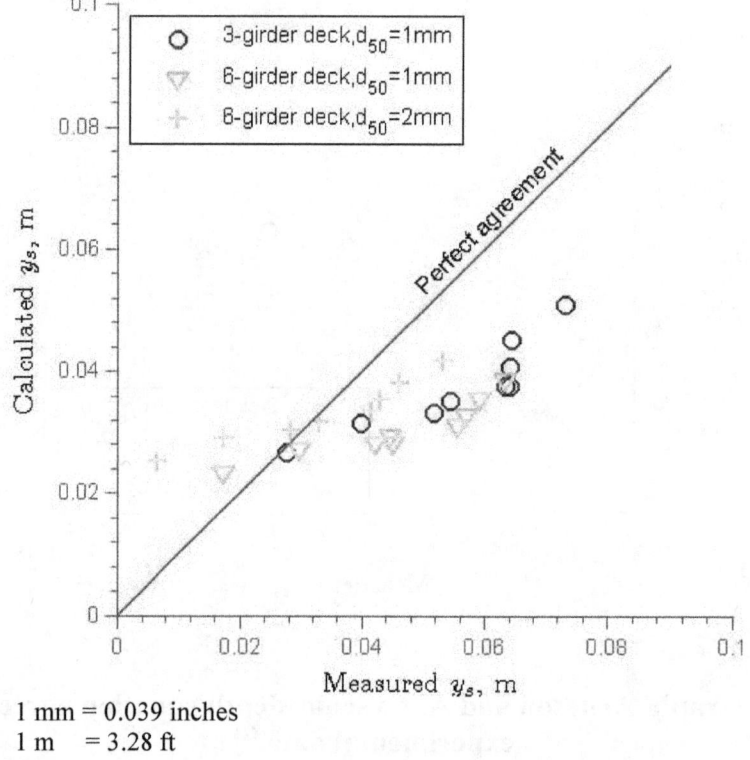

Figure 36. Graph. Umbrell et al.'s scour depth equation agreement with experimental data.[3]

CHAPTER 4. ANALYTICAL STUDY OF MAXIMUM SCOUR DEPTH

The experiments were conducted under two critical velocity conditions, but the purpose of the experiments was to make the results as widely applicable as possible. To achieve this end, an analytical solution for the maximum scour depth needed to be found.

For clarification, the problem is stated as follows:

Given a bridge over a steady river flow with clear water without contraction channel and piers (shown in plan view in figure 37) experiencing the flow conditions in either figure 38 through figure 40), find the equilibrium maximum pressure flow scour depth, y_s, per unit of river flow.

Where:

V_{uc} = Upstream critical velocity.
B = Width of the river.
W = Width of the bridge deck.
d_{50} = Median diameter of the bed materials.
h_u = Depth of the headwater.
h_b = Bridge opening before scour.
h_d = Depth of the tailwater.
b = Thickness of the bridge deck including girders.

FLOW CLASSIFICATION

The solution to the problem depends on the tailwater surface elevation. As in Picek et al., the bridge flows are divided into three cases.[6]

Case 1

If the downstream low chord of a bridge is unsubmerged as shown in figure 38, the bridge operates as an inlet control sluice gate. The scour is independent of the bridge width and continues until a uniform flow and a critical bed shear stress are reached. This case occurs only for upstream slightly submerged conditions. Since the flow condition under the bridge in this case is an open channel flow, it is presented in appendix A.

Case 2

If the downstream low chord is partially submerged as shown in figure 39, the bridge operates as an outlet control orifice, and the bridge flow is rapidly varied pressure flow.

Case 3

If the bridge is totally submerged as shown in figure 40, it operates as a combination of an orifice and a weir. Only the discharge through the bridge affects scour depth. In the following sections, only the solutions for cases 2 and 3 are discussed.

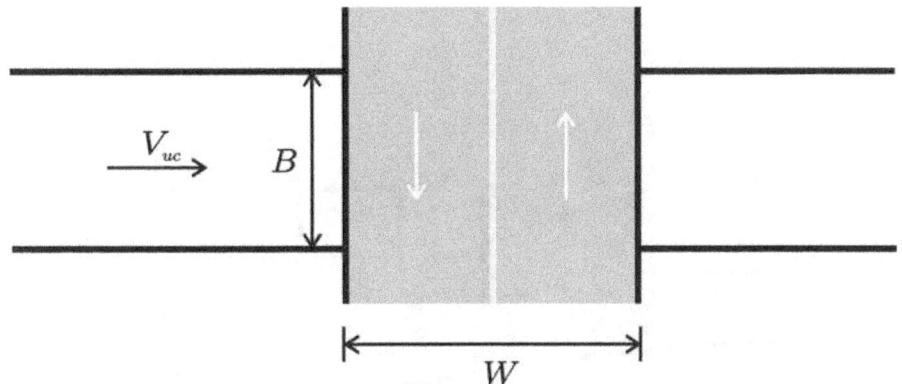

Figure 37. Illustration. Plan view of bridge over stream.

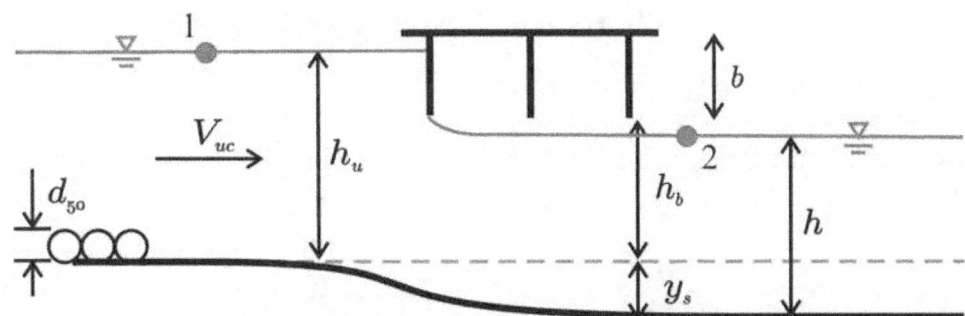

Figure 38. Illustration. Pressure flow for case 1.

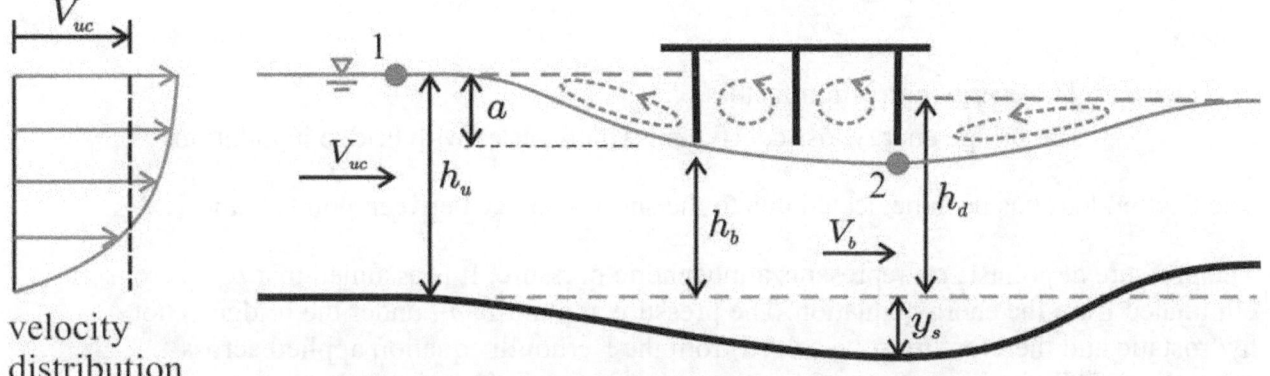

Figure 39. Illustration. Pressure flow for case 2.

23

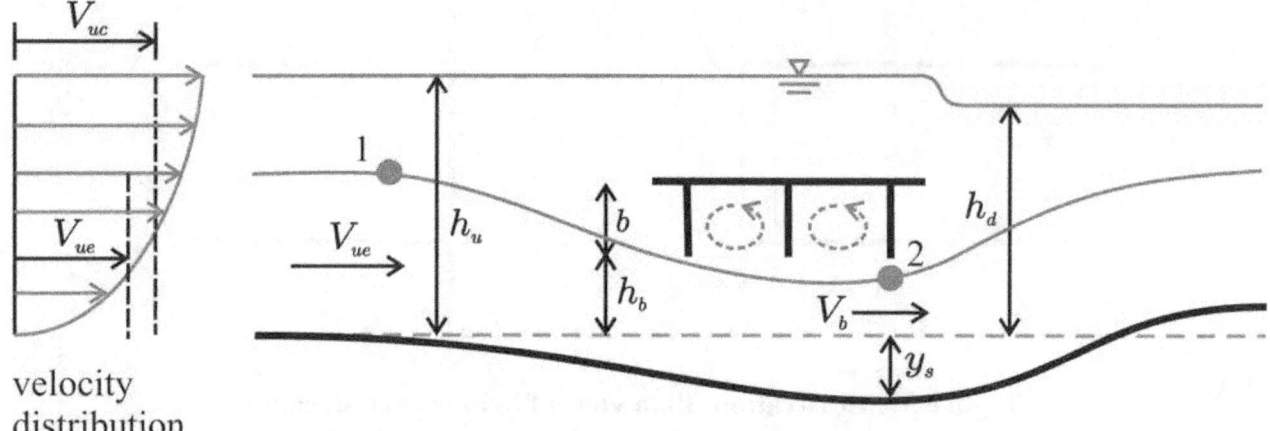

velocity
distribution

Figure 40. Illustration. Pressure flow for case 3.

CASE 2—PARTIALLY SUBMERGED FLOWS

Neglecting the friction, the energy equation along the streamline 1-2 is shown in the following equation in figure 41:

$$h_u + \frac{\alpha_1 V_{uc}^2}{2g} = h_b + \frac{p_2}{\gamma} + \frac{\alpha_2 V_b^2}{2g} + K_b \frac{V_b^2}{2g}$$

Figure 41. Equation. Energy equation along streamline 1-2.

Where:

α_1 and α_2 = The energy correction factors.
K_b = The bridge energy loss coefficient which varies with bridge inundation.

The friction loss has been neglected due to the short distance between points 1 and 2.

The pressure at point 1, p_1, represents atmospheric pressure. It is assumed that p_1 = zero, so it is eliminated from the energy equation. The pressure at point 2, p_2, under the bridge is not hydrostatic and therefore must be solved from the Bernoulli equation applied across streamlines.[9] Referring to figure 94 in appendix B, figure 42 is generated.

$$\frac{p_2}{\gamma} = (h_d - h_b) - K_p \frac{V_b^2}{2g}$$

Figure 42. Equation. Pressure under the bridge, p_2.

Where:

h_d = The depth of the tailwater.
K_p = A curvature coefficient that, like K_b, varies with bridge inundation.

The difference term in parentheses in figure 42 is the hydrostatic pressure head, while the last term is a curvature pressure head. Substituting the equation in figure 42 into figure 41 gives the following equation in figure 43:

$$h_u + \frac{\alpha_1 V_{uc}^2}{2g} = h_d + \frac{\alpha_2 V_b^2}{2g} + \left(K_b - K_p \right) \frac{V_b^2}{2g}$$

Figure 43. Equation. Energy equation including curvature coefficient.

Since both K_b and K_p are related to bridge inundation and must be zero when $h_u = h_b$ corresponds to open channel flow, the following equation in figure 44 is assumed:

$$K_b - K_p = \lambda \left[\frac{g(h_u - h_b)}{V_{uc}^2} \right]^{m/2}$$

Figure 44. Equation. Model describing difference bridge energy loss coefficient and curvature coefficient.

In figure 44, λ and m are determined experimentally. The gravitational acceleration, g, and V_{uc} are involved because of dimensional homogeneity. Substituting figure 44 into figure 43 and rearranging it gives the following equation in figure 45:

$$h_u + \frac{\alpha_1 V_{uc}^2}{2g} = h_d + \left\{ \alpha_2 + \lambda \left[\frac{g(h_u - h_b)}{V_{uc}^2} \right]^{m/2} \right\} \frac{V_b^2}{2g}$$

Figure 45. Equation. Energy equation including empirical parameters.

Figure 45 can be rearranged, as in the following equation in figure 46:

$$\frac{2g(h_u - h_d)}{V_{uc}^2} + \alpha_1 = \left\{ \alpha_2 + \lambda \left[\frac{g(h_u - h_b)}{V_{uc}^2} \right]^{m/2} \right\} \left(\frac{V_b}{V_{uc}} \right)^2$$

Figure 46. Equation. Rearrangement of energy equation including empirical parameters.

Considering the continuity in figure 47, as follows:

$$V_{uc}(h_b + a) = V_b(h_b + y_s)$$

Figure 47. Equation. Continuity equation.

Where:

$a = h_u - h_b$, as shown in figure 39.

The equation in figure 46 becomes figure 48, where the left side of the equation, $(h_b+y_s)/(h_b+a)$, is called the scour number.

$$\frac{h_b + y_s}{h_b + a} = \sqrt{\frac{\alpha_2 + \lambda\left[\dfrac{g(h_u - h_b)}{V_{uc}^2}\right]^{m/2}}{\alpha_1 + \dfrac{2g(h_u - h_d)}{V_{uc}^2}}}$$

Figure 48. Equation. Pressure flow scour design equation.

Unfortunately, the downstream flow depth, h_d, is unknown. For approximation, figure 49 is assumed to be a function defined as follows:

$$h_u - h_d = \beta\left(h_u - h_b\right)$$

Figure 49. Equation. Downstream flow depth approximation.

Where zero $< \beta < 1$. Concisely, F is defined in figure 50 as follows:

$$F = \frac{V_{uc}}{\sqrt{g\left(h_u - h_b\right)}}$$

Figure 50. Equation. Inundation Froude number.

Figure 50, when combined with the equations in figure 49 and figure 48, is reduced to the following equation in figure 51:

$$\frac{h_b + y_s}{h_b + a} = \sqrt{\frac{1 + \dfrac{\lambda}{F^m}}{1 + \dfrac{2\beta}{F^2}}}$$

Figure 51. Equation. Pressure flow scour design equation including inundation Froude number.

The values of α_1 and α_2 have been taken as 1, and the parameters λ, m, and β are determined experimentally. The equation in figure 51 will be tested after case 3 is discussed.

CASE 3—TOTALLY SUBMERGED FLOW

The solution for case 2 can be adapted to case 3 after a slight modification. This can be proven by applying the energy equation (figure 41) to the situation in figure 40. The effective velocity, V_{ue}, at point 1 is significantly affected by the bridge deck. As an approximation, the relation in the following equation in figure 52 is assumed:

$$\frac{V_{ue}}{V_{uc}} = \left(\frac{h_b + b}{h_u}\right)^{0.85}$$

Figure 52. Equation. Effective velocity equation.

The exponent 0.85 is a fitting constant derived from the data in the graph shown in figure 53.

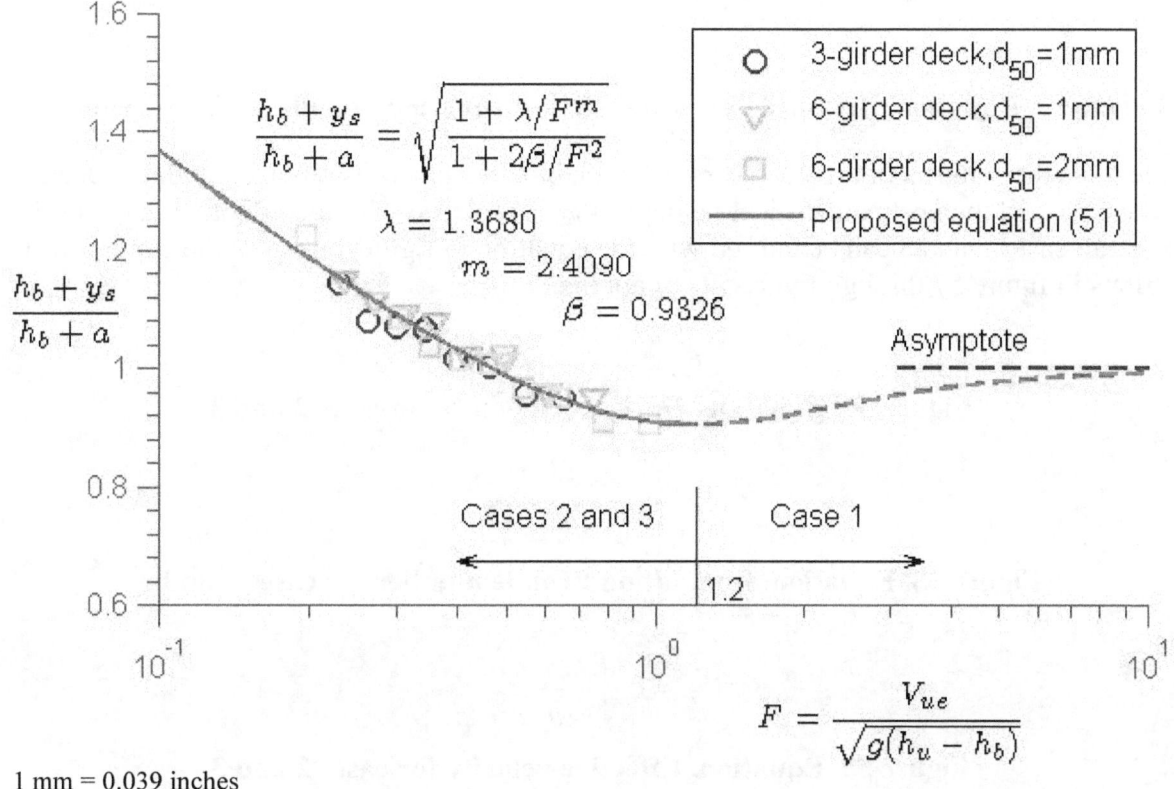

1 mm = 0.039 inches

Figure 53. Graph. Scour number versus inundation Froude number.

The unit discharge, q_1, through the bridge is then described in figure 54 as follows:

$$q_1 = (h_b + b)V_{ue}$$

Figure 54. Equation. Unit discharge.

27

The velocity at the maximum scour section is illustrated in the equation in figure 55 as follows:

$$V_b = \frac{q_1}{h_b + y_s} = \frac{h_b + b}{h_b + y_s} V_{ue}$$

Figure 55. Equation. Velocity at maximum scour section.

When the equation in figure 41 is applied to the situation in figure 40, the pressure at point 1, p_1, is hydrostatic, and the pressure at point 2, p_2, is the same as that in figure 42. Substituting the equation in figure 55 into figure 46 and rearranging it gives the following equation in figure 56:

$$\frac{h_b + y_s}{h_b + b} = \sqrt{\frac{1 + \lambda \left[\dfrac{g(h_u - h_b)}{V_{ue}^2} \right]^{m/2}}{1 + \dfrac{2g(h_u - h_d)}{V_{ue}^2}}}$$

Figure 56. Equation. Pressure flow scour design equation including effective velocity.

Figure 56 is the same as figure 48 except the deck block depth, a, is replaced with the deck thickness, b, and the upstream critical velocity, V_{uc}, is replaced with the effective velocity, V_{ue}. In general, cases 2 and 3 can be unified with the equation in figure 51 where the conditions in the equations in figure 57 through figure 59 are applied.

$$a = \min\{h_u - h_b, b\}$$

Figure 57. Equation. Deck block depth for cases 2 and 3.

$$F = \frac{V_{ue}}{\sqrt{g(h_u - h_b)}}$$

Figure 58. Equation. Inundation Froude number for cases 2 and 3.

$$V_{ue} = V_{uc} \left(\frac{h_b + a}{h_u} \right)^{0.85}$$

Figure 59. Equation. Effective velocity for cases 2 and 3.

Note that for case 2, the effective velocity, V_{ue}, reduces to the critical velocity, V_{uc}.

MAXIMUM SCOUR DEPTH

Test of the Hypothesis

It is hypothesized that the maximum scour depth for cases 2 and 3 can be described in figure 51 where λ and m are positive and zero $< \beta < 1$. To test the equation in figure 51, the inundation Froude number, F, and the scour number, $(h_b+y_s)/(h_b+a)$, for the experimental data are listed in table 1 through table 3 in columns 4 and 5, respectively, which are also plotted in figure 53. Applying the data to the equation in figure 51 and using the least-squares fitting function in MatLab®, the model parameters are obtained as follows:

Where:

λ = 1.3680.
m = 2.4090.
β = 0.9325.

The correlation coefficient $R^2 = 0.9639$.

Figure 53 shows that F and the scour number are appropriate similarity numbers to describe bridge pressure flow scour since all the data fall into a single curve regardless of bridge girder and sediment size. In addition, the curve has a minimum value at $F = 1.2$ and $(h_b+y_s)/(h_b+a) = 0.9055$, corresponding to the criterion between cases 1 and 2. The figure also shows that the proposed equation agrees well with the data when $0.2 \leq F \leq 1$, which corresponds to $1.14 \leq h_u/h_b \leq 3.57$. Finally, the dashed line for case 1 is an extension of figure 51. Mathematically, when h_u approaches h_b, F approaches infinity, and the scour number has an asymptote, $(h_b+y_s)/(h_b+a) \rightarrow 1$, which gives $y_s \rightarrow$ zero, since $a \rightarrow$ zero. This asymptote shows that the structure of figure 51 is reasonable. In terms of design, case 1 (where $F > 1.2$) is trivial since its scour is less than those of cases 2 and 3.

The Effect of Sediment Size

When sediment size increases, V_{uc} increases. The increase in V_{ue} can be computed according to the equation in figure 59. F then increases, which results in a decrease in the scour number. As a result, scour depth decreases with increasing sediment size.

The Effect of Deck Inundation

The bridge opening, h_b, appears in both axes in figure 53. To study the effect of h_b, the equation in figure 51 is rewritten in the equation in figure 60 as follows:

$$y_s = (h_b + a) \sqrt{\frac{1 + \dfrac{\lambda}{F^m}}{1 + \dfrac{2\beta}{F^2}}} - h_b$$

Figure 60. Equation. Maximum scour depth calculation.

For example, examine the six-girder deck with 0.078 inches (2 mm) of sediment for an experiment. The equation in figure 60 is plotted along with the measured experimental data in 1 mm = 0.039 inches
1 m   = 3.28 ft

figure 61, which shows that when $h_b > 0.164$ ft (0.05 m), scour depth decreases with increases in the bridge opening h_b. However, if $h_b \leq 0.164$ ft (0.05 m) or the deck is close to the bed, the scour calculation from figure 60 is significantly less than the measured value. This deviation results from the velocity profile near the bed. When the deck is close to the bed, V_{ue} is significantly smaller, and the energy correction factors α_1 and α_2 are much larger than the assumed value of 1. In other words, the proposed equation in figure 51 is only valid when the bridge deck is a sufficient distance above the bed, such as $h_b/h_u > 0.28$.

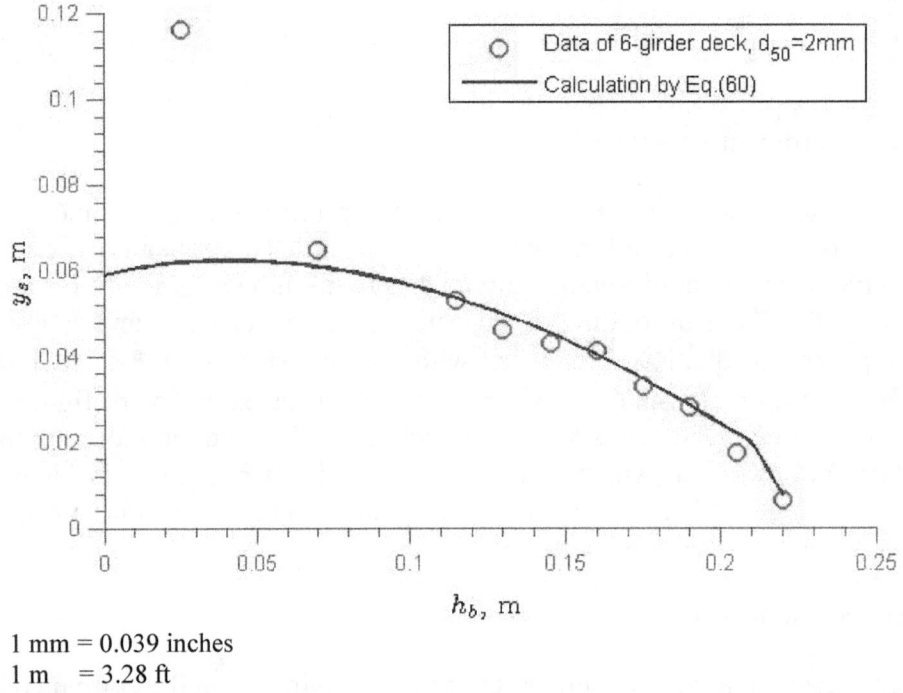

1 mm = 0.039 inches
1 m   = 3.28 ft

Figure 61. Graph. Maximum scour depth versus bridge opening height.

The Effect of Deck Thickness

The definition of deck thickness, b, is shown in figure 38 through figure 40. Obviously, for
case 2, the scour is independent of deck thickness. Nevertheless, for case 3, the scour varies with deck thickness. 1 mm = 0.039 inches
1 m   = 3.28 ft

Figure 62 shows an example of the y_s versus b relationship, assuming that all other variables remain constant. It shows that for case 3, y_s increases almost linearly with b. This implies that to reduce y_s, b should be minimized in design.

30

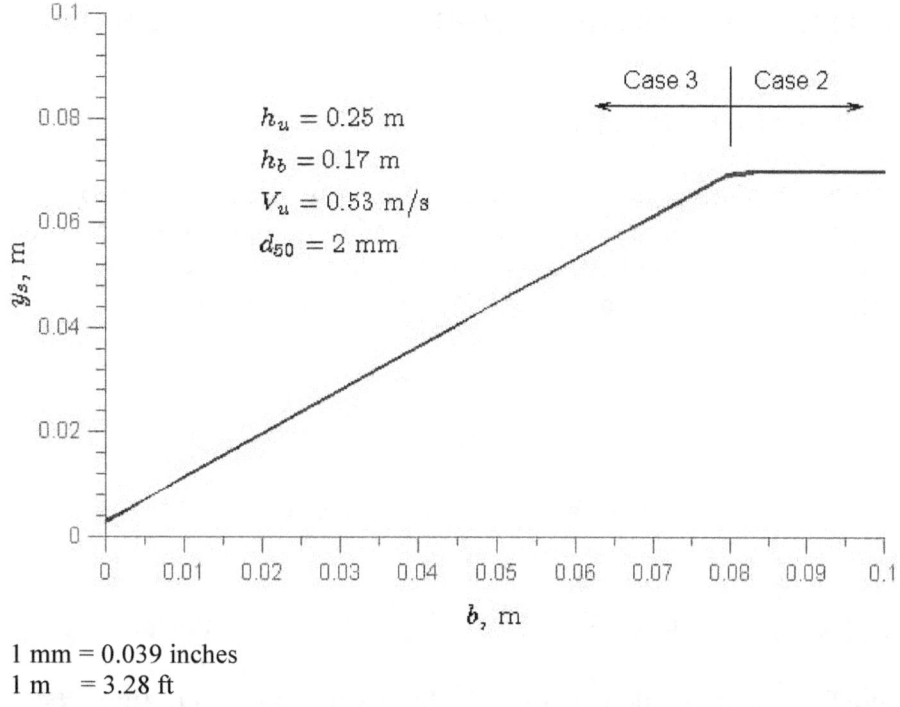

$h_u = 0.25$ m

$h_b = 0.17$ m

$V_u = 0.53$ m/s

$d_{50} = 2$ mm

1 mm = 0.039 inches
1 m = 3.28 ft

Figure 62. Graph. Maximum scour depth versus bridge thickness.

From this chapter, it is concluded that (1) case 2 or 3 occurs when $F \leq 1.2$; (2) cases 2 and 3 can be unified by the equation in figure 51 when the conditions in figure 57 through figure 59 are applied; and (3) once the maximum scour depth is estimated using figure 51, the scour profile can be calculated by the equations in figure 22 and figure 23. The next chapter focuses on the application of the results of this study through several examples.

CHAPTER 5. DESIGN PROCEDURE AND APPLICATION EXAMPLES

CRITICAL VELOCITY EQUATION

The design with the equation in figure 51 or figure 53 requires the critical approach velocity. Besides Neill's critical velocity equation and the equation in figure 4, several other velocity equations are available in the literature corresponding to a specific sediment size range. A general equation for the velocity can be derived from the Manning equation and the Shields diagram.[4]

$$V_c = \frac{\sqrt{K_s(s-1)d_{50}}}{n} h_u^{1/6}$$

Figure 63. Equation. Critical velocity.

Where:

K_s = The critical Shields number.

The Manning coefficient, n, is calculated via the following equation in figure 64:

$$n = 0.04(1.25d_{50})^{1/6}$$

Figure 64. Equation. Manning coefficient.

Substituting figure 64 into figure 63 gives the critical velocity equation in figure 65 as follows:

$$V_c = 7.69\sqrt{K_s(s-1)gd_{50}} \left(\frac{h_u}{d_{50}}\right)^{1/6}$$

Figure 65. Equation. Critical velocity.

In the equation in figure 65, the gravitational acceleration, g, is considered for dimensional homogeneity, and K_s can be approximated by figure 80 in appendix A.

Figure 4 is a special case of figure 65 where $K_s = 0.039$, corresponding to a sand diameter $d_{50} = 0.0585$ inches (1.5 mm). The equation in figure 65 is a general critical velocity equation for sands based on the Shields diagram, and it is recommended in this report.

Design Procedure

Consider the design procedure for the following problem:

Given a design unit discharge q, bridge opening h_b, deck thickness b, and bed material diameter d_{50}, find the scour depth, y_s, and scour profile.

The design procedure exists as follows:

1. Use the Hydrologic Engineering Centers River Analysis System (HEC-RAS) program to estimate the approach flow depth h_u. Note that the proposed method is based on rectangular flume experiments. For natural channels where flow depths are not uniform in the lateral dimension, a representative local depth, h_u, should be used.

2. Calculate the critical velocity, V_{uc}, from figure 65. If the upstream velocity, V_u, is less than or equal to V_{uc}, the proposed equation in this report is used. Otherwise, a procedure for live bed scour should be used.

3. Calculate the deck block depth, a, for clear water scour using the equation in figure 57.

4. Calculate the effective upstream velocity, V_{ue}, using the equation in figure 59.

5. Calculate the inundation Froude number, F, using the equation in figure 58 and check if the bridge flow is under pressure flow where $F \leq 1.2$.

6. Calculate the pressure flow scour depth, y_s, using the equation in figure 60.

7. Plot the design scour profile according to figure 22 and figure 23.

Column 6 in table 1 through table 3 is obtained using the above procedure. Column 7 shows that except for a few tests with little scour and in instances where the deck was positioned very close to the bed, most of the calculations generated using the equation in figure 51 are within 10 percent of the measured values.

APPLICATION EXAMPLES

Example 1 (Foundation Design)

The following example is modified from HEC-18:[4]

An existing bridge with a deck width of 30.28 ft (10 m) supported by five girders is subjected to pressure flow during a 100-year flood. There is only a small increase in flow depth at the bridge for the 500-year flood due to the large overbank area. The bed materials are characterized by a size of $d_{50} = 0.078$ inches (2 mm), and the bridge opening is $h_b = 26.01$ ft (7.93 m) before scour occurs. Assuming that the deck thickness including the girders and guardrail is $b = 6.56$ ft (2 m) for case 2 and $b = 3.28$ ft (1 m) for case 3, calculate the maximum vertical contraction scour depth and scour profile using the previously listed steps.

1. Assume that the HEC-RAS program is used to get the following flow conditions:

$h_b = 31.98$ ft (9.75 m).
$V_u = 3.28$ ft/s (1.0 m/s).
$q = 104.95$ ft^2/s (9.75 m^2/s).

2. Calculate the critical velocity according to figure 65. First, d_* and K_s (which are defined in figure 81 and figure 82) are calculated in figure 66.

$$d_* = \left[\frac{(s-1)g}{v^2}\right]^{1/3} d_{50} = \left[\frac{(1.65)(9.81)}{(10^{-6})^2}\right]^{1/3} (0.002) = 50.59$$

Figure 66. Equation. Dimensionless diameter.

The kinematic viscosity has been taken as $v = 10^{-6}$. K_s is then calculated as in figure 67.

$$K_s = \frac{0.23}{50.59} + 0.054\left[1 - \exp\left(-\frac{(50.59)^{0.85}}{23}\right)\right] = 0.0426$$

Figure 67. Equation. Critical Shields number.

V_{uc} is then calculated using figure 68.

$$V_{uc} = 7.69\sqrt{(0.0426)(1.65)(9.81)(0.002)}\left(\frac{9.75}{0.002}\right)^{1/6} = 1.176\,\text{m/s} > V_u$$

1 m = 3.28 ft

Figure 68. Equation. Critical approach velocity.

The results indicate that this is a clear water scour condition. For bridge safety, the critical velocity is applied.

3. Calculate the deck block depth for $b = 6.56$ ft (2 m), as seen in figure 69.

$$a = \min\{h_u - h_b, b\} = \min\{9.75 - 7.93, 2.0\} = 1.82\,\text{m}$$

1 m = 3.28 ft

Figure 69. Equation. Deck block depth evaluation.

The calculated deck block depth indicates that the flow is of the type represented by case 2. For case 2, the effective upstream velocity is the same as the upstream critical velocity, $V_{ue} = 3.86$ ft/s (1.176 m/s).

4. Calculate F, as shown in figure 70.

$$F = \frac{1.176}{\sqrt{(9.81)(9.75 - 7.93)}} = 0.2784 < 1.2$$

Figure 70. Equation. Inundation Froude number evaluation to determine pressure flow.

The results from figure 70 (i.e. $F < 1.2$) show the bridge flow is under a pressure flow condition, and can be described as either case 2 or 3.

Calculating the scour depth using the equation in figure 60, the results are shown in figure 71.

$$y_s = (7.93 + 1.82)\sqrt{\dfrac{1 + \dfrac{1.3680}{0.2784^{2.4090}}}{1 + \dfrac{2(0.9326)}{0.2784^2}}} - 7.93 = 2.874\,\text{m}$$

1 m = 3.28 ft

Figure 71. Equation. Scour depth evaluation.

From figure 30, y_s is at a distance of x_2 from the downstream deck edge. This distance is solved in 1 m = 3.28 ft

figure 72.

$$x_2 = 0.152W = (0.152)(10) = 1.52\,\text{m}$$

1 m = 3.28 ft

Figure 72. Equation. Maximum scour depth position.

5. Estimate the equilibrium scour profile, y, by the equations in figure 22 and figure 23, which is solved in figure 73 for the case when $x \leq$ zero.

$$y = -2.874\exp\left(-\left|\frac{x}{10}\right|^{2.5}\right)$$

Figure 73. Equation. Equilibrium scour profile equation, x is less than or equal to zero.

6. Calculate figure 74 for $x >$ zero.

$$y = -(2.874)(1.055)\exp\left[-\frac{1}{2}\left(\frac{x}{10}\right)^{1.8}\right] + (0.055)(2.874)$$

Figure 74. Equation. Equilibrium scour profile equation, x is greater than zero.

7. Simplify figure 74 to generate the equilibrium scour profile in figure 75.

$$y = -3.0321\exp\left[-\frac{1}{2}\left(\frac{x}{10}\right)^{1.8}\right] + 0.15807$$

Figure 75. Equation. Simplified equilibrium scour profile equation, x is greater than zero.

The equilibrium scour profile generated by figure 75 is plotted in   1 m = 3.28 ft
figure 76.

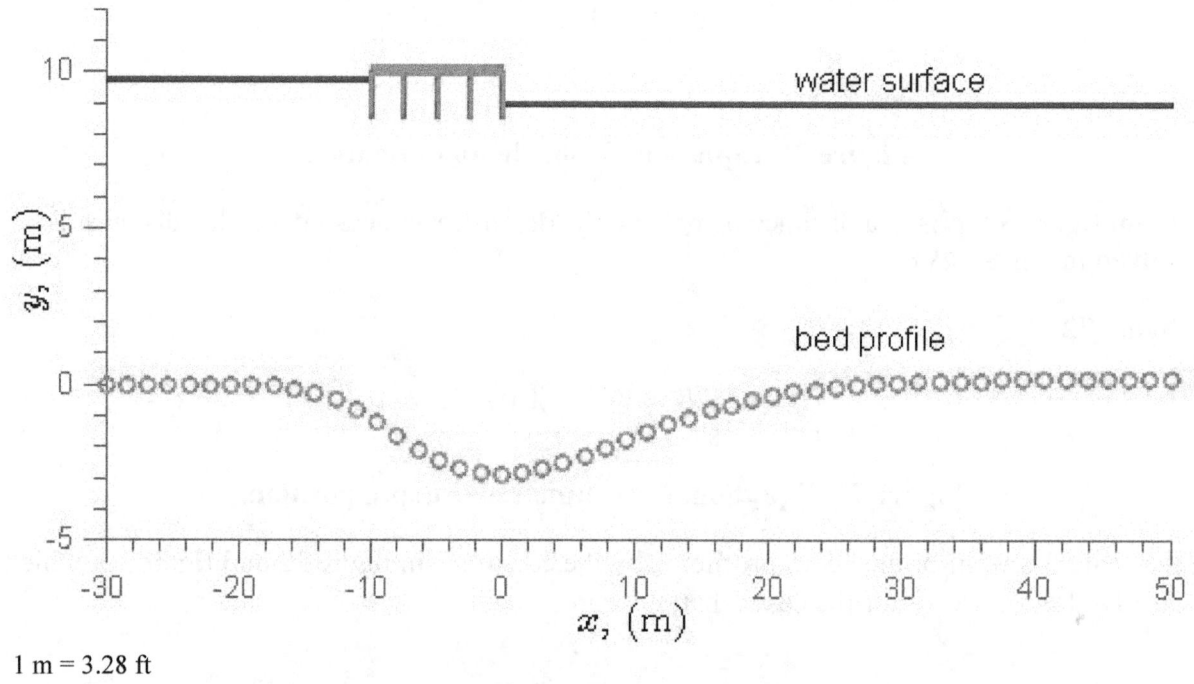

1 m = 3.28 ft

Figure 76. Graph. Scour profile for example problem.

Repeating the above steps with $b = 3.28$ ft (1 m), the maximum scour depth for case 3 can be found. This depth, as calculated by this method, and the other three methods examined previously is shown in the last column of table 4.

The maximum scour depths calculated according to the different methods are summarized in table 4. In general, the proposed method gives results in the same order of magnitude as the previous methods. Nevertheless, the results of the previous methods from the literature might be too conservative according to practical experience.

Table 4. Maximum scour depth estimates by four different methods.

Method	Maximum Scour Depth for Case 2, $b = 2$ m	Maximum Scour Depth for Case 3, $b = 1$ m
Arneson and Abt, figure 3	6.58	6.10
Lyn, figure 5	4.88	4.88
Umbrell et al., figure 7	8.29	9.52
Proposed method, figure 60	2.87	2.12

1 m = 3.28 ft

Example 2 (Scour Evaluation)

For a field scour evaluation for the previous example, if the scour depth is measured at the upstream deck edge, it is about -4.89 ft (-1.49 m), as seen in 1 m = 3.28 ft

figure 77.

$$y\big|_{\text{upstream deck edge}} = -1.49\,\text{m}$$
1 m = 3.28 ft

Figure 77. Equation. Scour depth at the upstream deck edge.

According to figure 32, the corresponding maximum y_s is 9.45 ft (2.88 m), as seen below in figure 78.

$$y_s = \frac{(-1.49)}{(-0.518)} = 2.88\,\text{m}$$
1 m = 3.28 ft

Figure 78. Equation. Maximum scour depth solution.

By comparing this scour depth with the designed foundation dimensions, a designer can determine whether or not the scour is critical and poses a risk to the structure.

CHAPTER 6. FURTHER RESEARCH NEEDS

This study is based on experiments in a rectangular flume using uniform sands with clear water at critical approach velocity and decks with rectangular girders that are far above the river bed. The results from the study are the maximum scour value and scour profiles. For cost efficiency, it would be beneficial to conduct additional research including the following:

- **Temporal variation of bridge pressure flow scour in clear water:** The methods proposed in this study are for computing equilibrium scour, which requires a long flow duration in flume experiments. The corresponding duration required to reach equilibrium scour in field conditions may be considerably larger than the duration of a design flood. Thus, further research is needed to estimate the variation in scour depth with respect to time.

- **Clear Water Scour with Approach Velocity Less Than Critical Velocity:** The proposed method assumes that the approach velocity is at critical velocity, which results in the maximum scour depth. When the approach velocity is less than the critical velocity, the scour should be smaller. Further research is needed to quantify this difference for various approach velocities.

- **Clear Water Scour with Sand Mixtures:** The proposed method is based on experiments using uniform bed materials. Heterogenous bed materials should have an armoring effect, which would reduce the scour depth in accordance with the median diameter. Further research is needed to quantify the effect of gradations in sediments.

- **Effect of Girder Shapes:** This study emphasized decks with rectangular girders. Although it was not described herein, a streamlined bridge deck shape was also preliminarily tested during the study. It showed a significantly shallower scour hole than those beneath the bridges with rectangular girders. Additional research is needed to study the effect of different deck and girder shapes on scour.

- **Pier Scour Under Bridge Pressure Flow Conditions:** The experiments in this study were conducted without piers. The current practice superposes a general scour and a local pier scour. This may not be a good assumption because of nonlinear interactions between a pier and a fluid under pressure flow conditions. Hence, further research is required to understand the nonlinear effects on pier scours under pressure flow conditions.

- **Live-Bed Scour in Bridge Pressure Flows:** The proposed method for clear water scour is governed by critical velocity. Nevertheless, most river flows are sediment-laden flows where scours are governed by sediment transport capacity. Hence, a study on live-bed scour is necessary.

Due to the limitations of the experiment documented in this study, engineering judgment should be exercised when developing new designs or retrofitting existing structures in natural channels using the proposed method.

CHAPTER 7. CONCLUSIONS

Several conclusions can be drawn from this study. A set of high-quality data on bridge pressure flow scour have been obtained. The data show that the horizontal scour depends on deck width. The scour starts at about 1 deck width upstream of the bridge, as shown in figure 28, and the deposition starts at about 2.5 deck widths downstream of the bridge, as shown in figure 31. A second conclusion is that a similarity scour profile exists where the horizontal length is normalized to the deck width and the vertical dimension to the maximum scour depth, as shown in figure 22 and figure 23. In addition, the similarity scour profile is mostly independent of the number bridge girders and sediment size. The dataset obtained can be a benchmark for further studies, and the similarity relations can be used for field scour evaluation.

An analytical solution for pressure flow scour has been presented. The theoretical study showed that the study of bridge flow scour can be divided into three cases: case 1 is open channel flow, while cases 2 and 3 are rapidly varied pressure flow. For pressure flow, the maximum scour depth can be described by a scour number and an inundation number, as in figure 51 or figure 53. The maximum scour depth decreases with increasing sediment size, but it increases with deck inundation and thickness. The analytical solution can predict the maximum scour and a corresponding scour profile. Since the analytical solution is based on the energy and mass conservation laws, it is expected to be applicable to prototype flows without scaling effects.

The proposed method has been validated with the flume data, and an application procedure with examples has been presented. Nevertheless, engineering judgment is required in practice when developing new designs or retrofitting existing structures in natural channels.

APPENDIX A. MAXIMUM SCOUR DEPTH FOR CASE 1

Referring to figure 38, when the scour reaches its equilibrium state, the downstream flow is uniform with a critical bed shear stress. If the uniform flow is described by the Manning equation and the critical bed shear stress by the Shields diagram, the downstream flow depth is the same as that in clear water contraction scour.[4]

$$h = \left[\frac{nq^2}{(s-1)d_{50}K_s} \right]^{3/7}$$

Figure 79. Equation. Downstream flow depth.

Where:

h = The downstream flow depth.
n = Manning coefficient.
q = Unit discharge.

The K_s for sands can be found with the equation in figure 80.[7]

$$K_s = \frac{0.23}{d_*} + 0.054 \left[1 - \exp\left(-\frac{d_*^{0.85}}{23} \right) \right]$$

Figure 80. Equation. Critical Shields number approximation by Guo.

In which:

$$K_s = \frac{\tau_c}{(s-1)\gamma d_{50}}$$

Figure 81. Equation. Shields number.

Where:

$$\tau_c =$$ The critical bed shear stress.

The dimensionless diameter, d_*, is defined in figure 82.

$$d_* = \left[\frac{(s-1)g}{\nu^2} \right]^{1/3} d_{50}$$

Figure 82. Equation. Dimensionless diameter.

Where:

v = The kinematic viscosity of water.

h = The downstream flow depth, which is the available uniform flow depth after scour.

The scour depth can be found by the energy equation between points 1 and 2 in figure 38 where the datum is chosen at the maximum scour bed elevation (see figure 83).

$$y_s + h_u + \frac{\alpha_1 V_u^2}{2g} = h + \frac{\alpha_2 V_b^2}{2g} + K_b \frac{V_b^2}{2g}$$

Figure 83. Equation. Energy equation between points 1 and 2.

Where:

α_1 and α_2 = Energy correction coefficients.

K_b = Entrance energy loss coefficient, which can be taken as 0.52 according to a box culvert experiment.[8] Note that the energy loss due to friction has been neglected because of the short distance between points 1 and 2.

The scour depth from figure 83 is then represented in figure 84 as follows:

$$y_s = h - h_u + \frac{q^2}{2gh^2}\left[\alpha_2 + K_b - \alpha_1\left(\frac{h}{h_u}\right)^2\right]$$

Figure 84. Equation. Scour depth.

In the figure, the relationship $V_u = q/h_u$ has been used. Theoretically, case 1 is well defined with figure 79 to figure 84. Practically, case 1 is only a short transition to case 2. This is because the upstream submerged portion of the bridge is not significant. As scour develops, the eroded materials will deposit somewhere downstream of the bridge. That sediment raises the tailwater and causes the downstream deck to become submerged.

APPENDIX B. DERIVATION OF PRESSURE UNDER BRIDGE DECK

The Bernoulli equation across streamlines is expressed below in figure 85.[9]

$$\frac{p}{\gamma} + z + \frac{1}{g}\int \frac{V^2}{R}\,dn = \text{constant across streamlines}$$

Figure 85. Equation. Bernoulli equation across streamlines.

Where:

R = Local radius of curvature of a streamline.
n = Normal coordinate to the streamline and toward concave side.

The flow through the maximum scour cross section can be simplified with circular streamlines and constant velocity, V_b, as shown in figure 87. Applying figure 85 to the vertical line gives figure 86 as follows:

$$\frac{p}{\gamma} + z + \frac{1}{g}\int_0^z \frac{V_b^2}{R_0 - z}\,dz = \text{const}$$

Figure 86. Equation. Bernoulli equation applied to circular streamlines.

The coordinates n and z are collinear along the vertical line that passes through the maximum scour point.

R_0 = The radius of curvature at the maximum scour point as shown in figure 81, and the local radius $R = R_0 - z$ at position z has been applied.

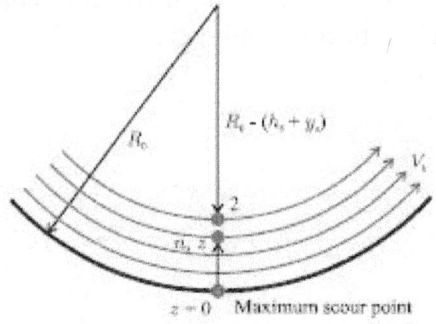

Figure 87. Illustration. Radii of curvature.

Integrating the equation in figure 86 gives figure 88.

$$\frac{p}{\gamma} + z - \frac{V_b^2}{g}\ln\frac{R_0 - z}{R_0} = \text{const}$$

Figure 88. Equation. Integration of figure 86.

44

Applying the equation in figure 88 to point 2 where $z_2 = h_b$ yields figure 89, which is valid for any velocity at point 2.

$$\frac{p_2}{\gamma} + h_b - \frac{V_b^2}{g} \ln \frac{R_0 - h_b}{R_0} = \text{const}$$

Figure 89. Equation. Bernoulli equation solved at point 2.

If V_b = zero from figure 39, the equation in figure 90 is generated as follows:

$$\frac{p_2}{\gamma} = h_d - h_b$$

Figure 90. Equation. Pressure at point 2 when V_b equals zero.

The downstream free surface is taken as the reference since it is close to point 2. Substituting figure 90 and V_b = zero into figure 89 gives the integration constant seen in figure 91 as follows:

$$\text{const} = h_d$$

Figure 91. Equation. Solution for integration constant.

Substituting figure 91 into figure 89 and rearranging it gives the general equation at point 2, as shown in figure 92 as follows:

$$\frac{p_2}{\gamma} = (h_d - h_b) + \frac{V_b^2}{g} \ln \left(1 - \frac{h_b}{R_0} \right)$$

Figure 92. Equation. Pressure at point 2.

The curvature coefficient, K_p, is defined in the equation figure 93 as follows:

$$K_p = -2 \ln \left(1 - \frac{h_b}{R_0} \right)$$

Figure 93. Equation. Curvature coefficient.

Through substitution, figure 92 becomes the equation seen in figure 94, in which the last term is called a curvature pressure. The parameter, K_p, represents the effect of the streamline curvature under the bridge. The equation in figure 93 is used in figure 41.

$$\frac{p_2}{\gamma} = (h_d - h_b) - K_p \frac{V_b^2}{2g}$$

Figure 94. Equation. Pressure at point 2 with curvature coefficient simplification.

ACKNOWLEDGEMENTS

This study was supported by the FHWA Hydraulics Research and Development Program with contract No. DTFH61-04-C-00037. We thank Oscar Berrios for his diligent and dedicated work in running the tests and preparing some of the figures. We are grateful to Mr. Bart Bergendahl at FHWA, Kevin Flora at the California Department of Transportation, and Professor Dennis Lyn at Purdue University for their constructive suggestions.

REFERENCES

1. Arneson, L.A. and Abt, S.R. (1998). "Vertical Contraction Scour at Bridges With Water Flowing Under Pressure Conditions," *Transportation Research Record 1647*, 10–17.

2. Lyn, D.A. (2008). "Pressure-Flow Scour: A Reexamination of the HEC-18 Equation," *Journal of Hydraulic Engineering, 134*(7), 1015–1020.

3. Umbrell, E.R., Young, G.K., Stein, S.M., and Jones, J.S. (1998). "Clear-Water Contraction Scour Under Bridges in Pressure Flow," *Journal of Hydraulic Engineering, 124*(2), 236–240.

4. Richardson, E.V. and Davis, S.R. (2001). *Evaluating Scour at Bridges: Fourth Edition*, HEC-18, FHWA-NHI 01-001, United States Department of Transportation, Washington, DC.

5. Neill, C.R. (1973). *Guide to Bridge Hydraulics*, University of Toronto Press, Toronto, Canada.

6. Picek, T., Havlik, A., Mattas, D., and Mares, K. (2007). "Hydraulic Calculation of Bridges at High Water Stages," *Journal of Hydraulic Research, 45*(3), 400–406.

7. Guo, J. (2002). Proceedings of the 13th IAHR-APD Congress, *World Scientific, 2*, 1096–1098.

8. Jones, J.S., Kerenyi, K., and Stein, S. (2006). *Effects of Inlet Geometry on Hydraulic Performance of Box Culverts*, FHWA-HRT-06-138, United States Department of Transportation, Washington, DC.

9. Young, D.F., Munson, B.R., Okiishi, T.H., and Huebsch, W.W. (2007). *A Brief Introduction to Fluid Mechanics*, 4th ed., John Wiley and Sons, Hoboken, NJ.